INTERACTIVE LEARNING AND
THE NEW TECHNOLOGIES

T0239533

COUNCIL OF EUROPE

INTERACTIVE LEARNING AND THE NEW TECHNOLOGIES

A REPORT OF THE EDUCATIONAL RESEARCH
WORKSHOP HELD IN EINDHOVEN (THE NETHERLANDS)
2-5 JUNE 1987

EDITED BY

COLIN HARRISON
UNIVERSITY OF NOTTINGHAM

Routledge
Taylor & Francis Group
LONDON AND NEW YORK

1988

First published 1989 by Routledge

2 Park Square, Milton Park, Abingdon, Oxfordshire OX14 4RN
52 Vanderbilt Avenue, New York, NY 10017

Routledge is an imprint of the Taylor & Francis Group, an informa business

First issued in paperback 2020

Library of Congress Cataloging-in-Publication Data

Educational Research Workshop (1987 : Eindhoven, Netherlands)
   Interactive learning and new technologies : a report of the
Educational Research Workshop held in Eindhoven, the Netherlands,
2-5 June 1987 / edited by Colin Harrison.
      p.   cm.
   Bibliography: p.
   ISBN 9026509693
   1. Computer-assisted instruction--Congresses. 2. Interactive
video--Congresses. 3. Educational technology--Congresses.
I. Harrison, Colin, 1945-   . II. Title.
LB1028.5.E335   1987
371.3'9446--dc19   88-36890 CIP

CIP-GEGEVENS KONINKLIJKE BIBLIOTHEEK, DEN HAAG

Interactive

Interactive learning and the new technologies : a report
of the educational research workshop held in Eindhoven
(the Netherlands), 2-5 June 1987 / ed. by Colin Harrison.
- Amsterdam [etc.] : Swets & Zeitlinger ; Berwyn : Swets
North America
Uitg. van Council of Europe. - Met lit. opg.
ISBN 90-265-0969-3 geb.
SISO 450.43 UDC 66/68: [331.5+371] NUGI 722
Trefw.: technologie en onderwijs / technologie en
arbeidsmarkt.

ISBN: 978-90-265-0969-8 (hbk)
ISBN: 978-0-367-60529-2 (pbk)

CONTENTS

PREFACE

The Eindhoven Workshop was one of a series of educational research meetings which have become an important element in the programme of the Council for Cultural Co-operation (CDCC) of the Council of Europe since 1975. European co-operation in educational research aims at providing Ministries of Education with research findings so as to enable them to prepare their policy decisions. Co-operation should also lead to a joint European evaluation of certain educational reforms and developments. The educational research meetings bring together research workers from the 25 countries taking part in the work of the CDCC. The purpose is to compare research findings on a particular topic of current interest; to identify areas of research so far neglected, and to discuss possibilities of research co-operation. The reports, as well as a selection of the papers of these meetings, are usually published as a book so that Ministries and research workers, as well as a wider public (teachers, parents, press) are kept informed of the present state of research in Europe.

The meeting in Eindhoven goes back to the suggestion made by the Dutch Ministry of Education and Science, and the Institute for Educational Research in the Netherlands (SVO), subsequently taken up by the CDCC. The Dutch Association for Scientific Film and Video (NVWFT) kindly agreed to organise the Workshop, in co-operation with the CDCC, the Dutch Ministry of Education and Science, and SVO. The Workshop was linked to a larger international conference called Media Manifestatie which dealt with interactive media in education and training. Both meetings took place at the Technical University of Eindhoven.

The theme "Interactive learning and the new technologies" was chosen because of the growing importance of the new technologies, both in education and the labour market. The idea was to put together research and experience to show in what way the new technologies had succeeded or can be successful in promoting classroom innovation through interactive learning.

Seven papers (covering France, the Federal Republic of Germany, Italy, the Netherlands, Sweden, and the United Kingdom) were commissioned; all but the Italian paper, the author of which was unable to attend, were presented in plenary session and then discussed in three working groups. National and individual reports from a number of countries, as well as lists of research projects and bibliographies, were tabled as background material. On the final day the Rapporteur General, Professor Dr. C.F. van der Klauw, from the Erasmus University of Rotterdam, summed up the discussion and conclusions.

1

The following countries were represented: Austria, Belgium, Denmark, Finland, France, the Federal Republic of Germany, Ireland, the Netherlands, Norway, Spain, Sweden, and the United Kingdom. There were also observers from the World Confederation of Organisations of the Teaching Profession, the International Federation of Secondary Teachers, the Association for International Curriculum Development and WORLDDIDAC. The list of participants is given at the end of this book.

The Council of Europe is particularly grateful to the Dutch Association for Scientific Film and Video (Dr. Jan Tijmen Goldschmeding and Drs. W.K. Sprij and their team), the Dutch Ministry of Education and Science (Dr. John J. de Wit and Mr. P. Morin), and the Institute for Educational Research in the Netherlands (Dr. J.G.L.C. Lodewijks and Drs. Rob Verkoeijen) for their excellent work in preparing and organising the Workshop. The Council of Europe would also like to express thanks to the Rapporteur General, Prof. Dr. C.F. van der Klauw, the lecturers, to the group chairmen and rapporteurs, and, last, but by no means least, to the editor, Dr. Colin Harrison.

Michael Vorbeck
Head of the Section for Educational
Research and Documentation

Strasbourg, 25 April 1988

2

PART 1:
REPORTS AND COMMISSIONED PAPERS

1.1 SECRETARIAT REPORT

by
the Section for Educational Research and Documentation
of the Council of Europe

1.1.1   OPENING

Dr. J.T. Goldschmeding, President of the Dutch Association of Scientific Film and Video (NVWFT) and Director of the Audiovisual Department of the Free University of Amsterdam, opened the meeting and welcomed the participants on behalf of the NVWFT, the Dutch Ministry of Education and Science, and the Institute for Educational Research in the Netherlands (SVO).

Mr. M. Vorbeck (Council of Europe) thanked NVWFT for having accepted the task of preparing and organising the Workshop in co-operation with the Council for Cultural Co-operation, the Ministry and SVO. He explained the nature of the work of the Council for Cultural Co-operation, and introduced the present workshop as part of a series of similar workshops held since 1975.

The meeting was chaired by Dr. J.T. Goldschmeding and others.

1.1.2   AIMS OF THE WORKSHOP

Mr. Vorbeck recalled the aims:

-   to take stock of and to evaluate research with regard to interactive learning and new technologies;

-   to draw attention to the consequences of research findings for educational policy and initial, as well as in-service education and training of teachers;

-   to identify areas of future research;

-   to encourage European research co-operation in this area.

1.1.3   ORGANISATION OF THE WORKSHOP

1.1.3.1   *Aspects discussed*

The workshop discussed aspects such as:

-   how best to bring about innovation in the classroom (which possibilities emerge from the new technologies to be adapted to classroom learning and teaching? Which combination of different media are thereby specifically suited for different classroom settings?)

5

- how to implement research results in classroom reality?

- how to create flexible school systems which might adapt to new situations?

- how best to organise the interplay between producers - research workers - teachers - teacher trainers - pupils, in developing software and learning material based on audiovisual aids.

- how best to familiarise teachers with these new developments, both in initial and in in-service education and training (INSET).

1.1.3.2   *Papers presented*

The following commissioned papers were presented in plenary session:

- Interactive learning and the new technologies; research into the use of computers for language and reading development in the United Kingdom, by Dr. Colin Harrison, University of Nottingham, United Kingdom;

- The Swedish way of implementing computer applications into the education of the nine-year compulsory school, by Dr. Göran Nydahl, Educational Software Group, Ministry of Education, Stockholm, Sweden;

- Aspects of the introduction of new technology into education in France: interactive learning and learning interaction, through research carried out at the INRP (National Institute for Educational Research, by Dr. Bernard Dumont, Director of the "New Technology and Education" Programme, Ministry of Education INRP, Montrouge, France;

- Interactive learning and new technologies in the Federal Republic of Germany, by Dr. Peter M. Fischer and Prof. Dr. Heinz Mandl, Deutsches Institut für Fernstudien an der Universität Tübingen, Federal Republic of Germany;

- Interactive learning with new technologies: when will it be successful? by P.W. Verhagen, University of Twente, The Netherlands;

- Computer-aided learning: a self-destroying prophecy? by Dr. Jef C.M.M. Moonen, Centre of Education and Information Technology (COI), University of Twente, The Netherlands.

1.1.3.3   *Background documents*

Individual or national research reports from Austria, Finland, the Federal Republic of Germany, Ireland, Italy, Norway, Portugal, Spain, the United Kingdom and the United States of America were tabled as background documents.

In addition, there were the following documents:

- A bibliography (DECS/Rech (87) 3);
- A list of ongoing or completed research compiled by the Secretariat on the basis of national contributions (DECS/Rech (87) 2);
- A "Reader", prepared by SVO and listing research projects and publications.

1.1.3.4 *Group work*

The commissioned papers were discussed in three groups. Their reports form section 1.2 of this book.

1.1.3.5 *Final report*

The Rapporteur General, Prof. Dr. C.F. van der Klauw, presented a verbal report.

1.1.4 SECRETARIAT SUMMARY

As for the Secretariat, the following conclusions might be drawn:

1.1.4.1. The newest technologies (e.g. interactive video, interactive compact disc, CD-ROM, two-way cable and satellite communication, viewdata and other remote database facilities) offer enormous possibilities to teachers and pupils, but the main problems remain software quality and teacher training. The earlier euphoria about the new technologies is fading away: expectations were too high.

1.1.4.2 The majority of teachers have never seen the new technologies in action and are therefore sceptical as regards their usefulness. To train them will be a long-term task.

1.1.4.3 Using the new technologies will require a new teaching style (e.g. a lot of group work); they will only be cost-effective if combined with peer teaching.

1.1.4.4 Software design will have to be multidisciplinary and be based on well-defined learning objectives and the felt needs of (the few convinced) teachers. A great deal of commercial software available does not meet these requirements, and therefore does not find its way into classrooms. Many commercial software producers have withdrawn from the field of education in the last 2-3 years. The governments may have to step in and fund development of prototype software.

1.1.4.5 Most programmes are neither good nor bad; much depends on how the teachers use them. Drill and practice programmes are declining in favour of more intelligent programmes diagnosing pupils' mistakes and offering advice and guidance. There is evidence that the new technologies have led to innovation and improvements in primary education (France, Netherlands, United Kingdom, United States of America), whereas - as a result of rigid curriculum structures - they have so far had little influence on secondary education.

1.1.4.6　Familiarity with new technologies will be useful in working life, but schools should abstain from training narrow computer specialists.

1.1.4.7　Several participants will in future co-operate in their research (e.g. France-Netherlands).

1.2 REPORTS OF WORKING GROUPS

1.2.1   REPORT OF GROUP A

Chairman: Dr. J.J. Beishuizen
Rapporteur: Dr. Colin Harrison

1.2.1.1   *The state of the art*

1.2.1.1.1   The French representative noted that besides national initiatives, regionalisation has gathered more and more importance in political as well as financial decisions. New technology developments have affected three main areas:

-    Informatics, following "Plan Information pour Tous", even if evaluation is not yet proposed, microcomputers are to be seen everywhere in primary schools and secondary schools and about 120,000 teachers have followed a week-long training course.

-    Audiovisual equipment: every college is going to be equipped with portable video equipment.

-    The increasing use of "telematics" (database, communications networks) through local computer networks as well as through regional or national ones (cf. Minitel).

1.2.1.1.2   From Spain the Group heard of an experimental and innovative New Information Technology (NIT) programme that provides equipment, training and funds for schools involved in its two main projects, Atenea and Mercurio. These projects were aimed at introducing computer and audiovisual technology in primary and secondary schools. Further developments were awaiting the results of education and reflection.

1.2.1.1.3   In the Netherlands there is a national 5 year development plan on the use of computers. However, this has been implemented within a decentralised system of education. Hardware was more widespread in high schools and further education, although it was felt that the decision to give in-service training to teachers at all levels has been a good one. There is also an experimental study with a small budget on interactive media.

1.2.1.1.4   The United Kingdom delegate reported a similar picture to the French one, although interestingly the initial major funding has been delivered via the Department of Trade and Industry, rather than the Department of Education and Science. Similarly, in Spain, innovation has been developed with the collaboration of government departments of education and industry.

1.2.1.1.5   The Belgian representative reported that there had been more research into microcomputers in primary schools, but more funding for secondary schools and further education.

1.2.1.1.6   In Finland, all teacher training institutions, all gymnasium schools and most comprehensive schools have computers. Provision is less in elementary schools.

1.2.1.2   *Problems of implementation*

1.2.1.2.1   A crucial point to emerge was that we cannot easily define a generalised model of implementation. Major differences in economic and educational factors in different European countries imply the need for a contextualised view of implementation of new technology.

1.2.1.2.2   For example, the Dutch and Finnish delegates emphasised the problem of:

   1.   difficulties in changing teachers' attitudes, and
   2.   integrating NIT into the curriculum.

1.2.1.2.3   By contrast French colleagues felt that there was a prior problem, namely a "scholastic" model of instruction in some countries, and that to attempt to graft NIT onto the existing model would be inappropriate and doomed to failure.

1.2.1.2.4   Many colleagues felt that the model of the teacher as a transmitter of knowledge was likely to need changing before NIT could be implemented successfully. A model in which the teacher was a mediator or facilitator would be more appropriate.

1.2.1.2.5   The importance of TIMING decisions relating to curriculum development was felt to be crucial. Poorly timed decisions could lead to poor hardware choices or ineffective in-service training.

1.2.1.3   *What research is needed?*

1.2.1.3.1   The prior question here is this: what models of research are appropriate and for what audience?

1.2.1.3.2   For civil servants and funding agencies data on cognitive gains are often sought and preferred.

1.2.1.3.3   For teachers, information on how a programma (or piece of hardware) will work in the classroom may be valuable. The group felt that in carrying out evaluation on NIT, teachers' judgements and observations of student activity should be taken into account.

1.2.1.3.4   However, it was pointed out that the fact that every bank, travel agent, telephone system, and designer in Europe now uses computers, presents in itself an urgent need for the introduction of NIT into schools.

1.2.1.3.5   It was felt that we still lack information concerning how computers are currently being used and whether innovation was occurring in many countries.

1.2.1.3.6   It was felt that research should be devoted to the effects of NIT on metacognition and self-reflection, and that these aspects should not be ignored by default.

1.2.1.3.7   The value of conducting integrated research into video and computers was stressed.

1.2.1.3.8   The group endorsed the importance of research into differences between the attitude and interest of boys and girls in new technology, especially microcomputers.

1.2.1.3.9   Research is needed into new modes of learning, such as the increasingly important role of images in the process of teaching and learning.

1.2.1.4   *Teacher training initiatives*

1.2.1.4.1   Pre-service training was considered to be a crucial factor in successfully integrating NIT in education. Without pre-service training NIT cannot serve as a change agent in education.

1.2.1.4.2   Both pre-service and in-service training should take account of the following:

-   orientation towards NIT should not stand on its own; it must be integrated within the curriculum as a whole;

-   input in training must come from pedagogy, educational psychology and computer science;

-   the focus should not be restricted to computers; an integrated approach towards all interactive technologies was advocated;

-   training should include attention to the criteria for evaluating software;

-   new technology has many applications in the wide world outside school;

-   programming should not be part of teacher training.

1.2.1.4.3   The cascade model was put forward as an example of one model of in-service training. However, special attention must be given to the problem of a disrupted flow of information. Teachers should be able to integrate their classroom experiences into in-service training. In-service training should be followed by a prolonged opportunity to exchange ideas and experiences.

1.2.1.4.4   In order to establish this opportunity, networks of teachers are advocated. These networks may serve as a communication channel between teachers who are involved in the NIT. They can exchange examples of successful and unsuccessful applications.

1.2.1.4.5   In order to establish and maintain appropriate levels of teacher motivation, a formal acknowledgement of having been successfully trained should be provided.

1.2.1.46   It is of particular importance to find adequate means to raise and sustain the interest of female teachers in this new technology.

1.2.2   REPORT OF GROUP B

Chairman: Dr. Jef C.M.M. Moonen
Rapporteur: Dr. David Owens

This report summarises the discussion under each of the five questions suggested by the Rapporteur General.

1.2.2.1   *The state of the art*

1.2.1.1.1   A disparity exists among the countries represented in the group concerning the amounts of money devoted by governments to support-ing the introduction of NIT in primary and secondary education. Where funds have been provided for the first stage of this process, funding is often insecure and not guaranteed for the next stages. Some governments, though committed in principle to the introduction of NIT, have provided only minimal funding for it. Most governments have underestimated the level of effort and funding required to install NIT.

1.2.2.1.2   If the majority of teachers are to be convinced of the value of NIT in the classroom, more research is needed on innovation strategies for implementing NIT.

1.2.2.1.3   Promoters of NIT in education should be cautious about claiming that it will lead to financial savings in primary and secondary education, and that it will lead to a reduction in the number of teachers.

1.2.2.1.4   It is unrealistic to expect teachers to become publishers or creators of much of their own computer software; they seldom publish their own textbooks. They need to be provided with tools to enable them to modify existing software.

1.2.2.2   *Problems of implementation*

1.2.2.2.1   The difficulties of European co-operation in software de-velopment were highlighted. We can co-operate by sharing ideas, tools for software development, and experiences with different types of user inter-faces. However, European courseware development is difficult because of cultural and social differences, and because historical experience of intro-ducing media into schools in each country differs.

1.2.2.2.2   A gap exists between researchers and the majority of teach-ers who are largely unaware of the potential of NIT in their classrooms. We need to bridge this gap by providing a new type of mediator who understands both the problems of the teachers and the concerns of the researchers, and is capable of the designing and development of software.

1.2.2.2.3   National differences exist in the attitude of governments towards industrial sponsorship of NIT in schools. The discussion group felt that industry was a lucrative potential source of funding.

1.2.2.3   *What research is needed?*

1.2.2.3.1   We should avoid spending the majority of research and development budgets on "cutting-edge" technologies; rather, we should focus on helping teachers make more effective use of "available" technology.

1.2.2.4   *Teacher training initiatives*

1.2.2.4.1   Knowledge, skills and opportunities for practical application are essential if teachers are to be able to use NIT in the classroom, therefore, pre-service and in-service training in NIT is crucial.

1.2.2.4.2   In-service courses should be organised within individual schools so that the unit of instruction is at the school. This is in order to be able to discuss the total impact of NIT at the school level.

1.2.2.4.3   NIT should be introduced throughout the curriculum, rather than as a separate curricular subject. Consequently, greater attention needs to be given to NIT in all courses in pre-service training.

1.2.2.5   *Summary*

Five major categories of research are required:

1.2.2.5.1   Laboratory or basic research - which does not need to have a direct link to practical application.

1.2.2.5.2   Prototype research - in which the possibilities of NIT are stressed to its limits and in which the most up-to-date research findings on learning, motivation, human factors and so on, are incorporated into software.

1.2.2.5.3   Practical applications research - which takes place in the classroom under genuine teaching conditions, for instance, as regards the impact of NIT on didactic methods within specific areas.

1.2.2.5.4   Innovation strategies - we need to clarify how we can speed up teacher acceptance of NIT by conducting research on innovation strategies.

1.2.2.5.5   "Future research" which informs what the contents of curricula should be, and what the potential roles of NIT might be.

1.2.3   REPORT OF GROUP C

Chairman: Mr. R.N. Tucker

1.2.3.1   *Point - Context*

Statement

The political, social educational, methodological context conditions what happened and can happen in each country and must be kept in mind.

Research proposal

Need better case studies with common descriptors if we are to compare experience or take parts of a model over to our own country.

1.2.3.2   *Point - Design*

Statement

New sorts of skill need to be brought together to specify the educational design. This needs government funding. Production and management skills need, for the production, commercial funding.

Because of the commercial difficulties for traditional publishers producing educational software, there is a need for government funding to establish specialised educational software producers.

Research proposal

At each state of development and testing the research findings feed back into the design. Equally, the market research and post-sales research which should be required of the commercial producers should be fed back to the designers of the next programs. Government and commercial funding are both needed for this to take place.

1.2.3.3   *Point - Role changes of teachers. Changes of methodologies*

Statement

The new media will change methods and roles. Different contexts are likely to produce different effects. There is a need for teachers to be trained and to accept these changes (or know how and why they should pre-empt them).

Research proposal

What are the new needs for the training and qualification of teachers? What is needed in terms of classroom management? What new evaluative and assessment skills are needed?

1.2.3.4   *Point - Primary schools*

Statement

Some countries find the greatest benefit from new media in primary schools, others have decided that there is no place for the new media in primary schools.

Research proposal

Need for research on the needs of primary schools followed by development of models from strictly controlled contexts.

1.2.3.5   *Point - Tools*

Statement

There appears to be a trend towards the use of utility/applications programs. Need to be a better description of what these are at the different levels.

Research proposal

Need description of the perceived merits and demerits of these programs to help in the selection of the right sort of tools for the right reasons and to plan any remedial action to compensate for any apparent demerits.

1.2.3.6   *Point - Networks*

Statement

Networks are seen as necessary for computer studies. Many were first bought because of cost of memory store. Now this is cheaper. What then are the reasons for having networks at classroom, school, inter-school (local) and regional level?

Research proposal

What networks and system extensions are available? What are the educational/operational merits and demerits? How can children best work together in groups/communicate with networks? How do they affect student/teacher/educational interactions?

1.2.3.7   *Point - INSET In-service ed & training*

Statement

There is an obvious need for introductory training and access, but also for steady support over a period of 3-5 years to create effective use and the required attitudinal change. This has long-term consequences.

Research proposal

What training and support is needed NOW and in the short-term? What training and support is needed in the longer term, and how can it be delivered?

1.2.3.8   *Point - Software toolkit and the computer-aided teacher (CAT)*

Statement

In order to provide the teacher with effective instruments to extend their abilities there should be a Toolkit of utility programs. See point 1.2.3.5. Producers should ensure that programs contain complete examples of use as demonstrations. For the teacher there should be further examples of how to teach with or through the programme. Where possible there should be case studies of effective classroom use. In this way we could use the technology to teach about the technology and ensure the provision of supportive self-study material to the teacher as a part of the INSET programme.

Research proposal

Developmental projects might be needed to create these INSET materials, but most of the skills already exist.

These will have to be evaluated/improved in each country.

1.2.3.9 *Point - Evaluation*

Statement

Evaluation of the use of new media in the classroom is linked to all these areas of development. It can be seen as formative, if used to improve the next stage of development.

Research proposal

New evaluation procedures may be necessary to measure what is happening in these more interactive learning environments, e.g. soft or qualitative methods.

1.2.3.10 *Point - Cost-effectiveness*

Statement

Is it possible to begin by determining the improvement in the learner and then calculate what costs should be attributed to that learning, rather than set up cost-effectiveness arguments based on comparisons of new and traditional media?

Research proposal

It would be useful to have scenarios based on the realities of school practice as a basis for such costing.

1.2.3.11 *Point - Computers NOT being used*

Statement

Several reports noted declining use other than in computer studies. Why? Partly this is a question of training. There is a need for more good descriptions of good and different uses of computers and video programmes in staff rooms, libraries and resource centres.

Research proposal

Need for research on what is the CRITICAL MASS in qualitative, quantitative and contextual terms to create this affective change.

1.2.3.12 *Point - Time / react now*

Statement

Moonen stated that the development needs time, and that we have to act now to avoid catastrophe. How do we cope with this dichotomy?

Research proposal

What do we need to do now and what needs to be done over a longer time? This needs good description with government support and agreement.

1.2.3.13 *Point - Integration*

Statement

Where do we begin reintegrating the curriculum to accommodate the new media. Many strategies could be used from the whole school approach to the support of the "good" teacher. The good experiences should then be disseminated, possibly through an "infective" model.

Research proposal

Need for good instructions of these strategies matched to the possibilities and limitations of each country's educational sub-contexts.

Note added by Group C:

Although not discussed at the Workshop we feel that there is need for more research into the knowledge processing or knowledge mapping that happen through or as a result of a child learning through these new media.

1.3 INTERACTIVE LEARNING AND THE NEW TECHNOLOGIES: RESEARCH INTO THE USE OF COMPUTERS FOR LANGUAGE AND READING DEVELOPMENT IN THE UNITED KINGDOM

by
Dr. COLIN HARRISON,
University of Nottingham

1.3.1 SUMMARY

In this paper it is argued that in the UK the emphasis on small group work using a microcomputer is partly the result of pedagogical opinion, and partly an economic necessity. The results of research studies into the best seating and grouping arrangements are discussed, and a number of evaluations of interactive programs for developing reading and language skills are reported. These focus on word processing, adventure games and programs used with slower learners. Aspects of the interaction between children, teachers and programs are discussed, and it is argued that the concept of an 'interaction continuum', which focusses attention on where a program lies on an intolerant-intelligent continuum, is a potentially useful one.

1.3.2 INTRODUCTION

The concept of interaction has been a central one in the development of thinking and practice in the introduction of computers into schools in the United Kingdom. In this introduction it will be argued that this has been so for two reasons, one pedagogic and the other economic. In the later sections of the paper we shall examine a number of research enquiries on the use of computers in the classroom which explore different aspects of interaction in the classroom.

The overal emphasis in the paper will be on the applications of microcomputers in the reading and language field, but in certain respects this emphasis could appear misleadingly narrow, since reading and language development have been the central objectives of much computer software in the UK, over a wide variety of school subjects, including mathematics and science.

The evidence that the concept of interaction is pervasive in British thinking about the use of computers is widespread. In its policy statement for members, the National Association for the Teaching of English (NATE, 1986) made it clear that its preferred teaching model was that of students working in small groups round a computer (see, in particular, paragraphs 20, 22, 26, 34, and 42). Some illustrative comments from the document will give some idea of the way in which the view is expressed:

> 'Where the words on the screen are agreed by a group, the technology can encourage exploration and risk taking ... Programs such as these can stimulate discussion ... this is a collective activity ... Computers lend themselves to small group work.'

One could add at this point that there is in general a correspondingly cool disposition in the NATE document towards drill and practice "skills" programs which teach such things as spelling or punctuation through essentially linear courses, and towards programs which are used as a substitute for interpersonal communication, and which might therefore be seen as leading to 'impoverished language' (paragraph 5).

In the broader area of Humanities teaching, the authors of the best known book in the field (Adams and Jones, 1983) make it clear in their introductory chapter that they hold a very negative view of software which aims at simple transmission of knowledge, and which tests the acquisition of that knowledge through multiple-choice questions. Equally, in their appendix on evaluation, it is clear that the question 'Is the content of educational value?' is closely related to a later one - 'Does the program allow for student interaction and/or creativity?'

A similar emphasis on interaction is to be found in the work of the largest UK project on the use of the computer in the teaching of mathematics (ITMA-Investigations on Teaching with Microcomputers as an Aid). In the project team's seminal book on program design and evaluation (Burkhardt et al, 1982), great emphasis is placed on the importance of observing the nature of the students' activity, which might include drawing, discussing, problem solving, experimenting, recording, discovering, interacting, explaining, reacting, interpreting, and story telling, as well as the more traditional calculating or learning.

In all these subject areas, therefore, the concept of interaction is taken to be a pedagogical imperative. It is, however, possible to argue that this undoubted emphasis on the importance of interaction in the UK is also a function of the country's economic state. The UK government made a significant investment in microcomputers in the period 1981-4, and funding from the Department of Industry enabled every school in the country to obtain one computer, and to buy others at half price. However, in late 1986 it was estimated that the ratio of computers to students in schools in England was no better than 1:95.

In this climate, the development of an educational philosophy which stressed the use of computers with groups rather than individuals was perhaps inevitable, and the emergent teaching strategies were as much the product of economic imperatives as the result of pedagogical debate.

1.3.3   INTERACTION IN ACTION - THE NATURE OF UK STUDIES

The remainder of this paper will address itself to the nature of children's interaction with computers in UK classrooms. As a coda, there will be a brief section on the types of response a computer can make, and the implications of these for developing students' learning.

There are many possible ways of evaluating the use of computer software. In an earlier paper (Harrison, 1985) I used Robert Stake's (1967) matrix of evaluation concerns to draw attention to the fact that, in the UK, both published checklists and research studies gave much greater emphasis to antecedents and transactions than to outcome. Equally, there was a much greater emphasis on judgements than on observations.

In this paper, therefore, a similar balance will obtain; a number of studies report learning outcomes, but in most the dependent variable is the nature or quality of interaction itself. As Hope (1985) has noted, if we begin by taking the view that it is the interaction between child, teacher, program and peers which is important, then certain things follow: if it is the quality of the interaction which determines "the value of the educational experience" then "life becomes more complex...". This is because "We can no longer talk about good and bad programs _per se_."

1.3.3.1   _Potter's 1985 study_

One of the most interesting and systematic studies of children working with computers on reading and language tasks is that of Frank Potter (1985). He reports a number of interrelated studies, all based on his work with a primary school teacher in one class of nine and ten year olds over a period of one school term. The number of children in the study was therefore not large, but the results are no less worthwhile for that.

The central problem posed by the study was to identify, if possible, the optimal conditions for undertaking group work using a microcomputer. The tasks in his programs involved reading a text on the screen, then undertaking some group activity, such as proposing, discussing and inserting new words to replace others which had been deleted by the teacher. His conclusions are based on careful observation, with time on task as the main dependent variable. In his preamble to the study, Potter mentions that his preference is for mixed ability groups, rather than groups of homogeneous ability. He quotes a study by Taylor (1985) which supports this approach.

Overall, Potter found that group work was most effective when it was preceded by a whole-class session of orientation, during which the children were shown the program in action, and offered a model of how to run it and respond to it. It was found that discussion became more full and fruitful towards the end of the task, and Potter therefore sought ways to maximise the effectiveness of the post-task review session. He found that it was better to delay review, and that the optimal period of delay was 24 hours (p<0.05). Even a short delay of 15 minutes seemed to be helpful in assisting the students to gain a little distance on the task of reviewing their answers, but a day's delay was most effective.

Potter also conducted an interesting study of the seating arrangements of the groups, which yielded statistically significant results. He attempted to facilitate discussion by seating the children, not in a row at a rectangular table, which was the ususal arrangement (see Figure 1a), but instead in a semicircle (Figure 1b). This new arrangement did require some preparation, since the BBC microcomputer does not have a long RGB lead, and a new one had to be made up and fitted, but the children preferred it (p<0.05). When interviewed at the end of term, most of the children said that in the semicircle they could see better, could talk to their peers more easily, could change places more readily, and could reach the keyboard more easily.

Another study compared group size, in order to determine the most satisfactory number to be working at the microcomputer. Potter compared three groups each of 3, 4 and 5 (i.e. nine groups in all), and he reports that he had expected 3 to be the best, but this was not the case. Both for indivi-

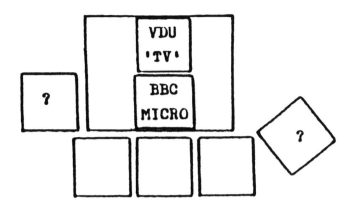

Figure 1a: Usual arrangement of seating round a microcomputer
               (Potter, 1985)

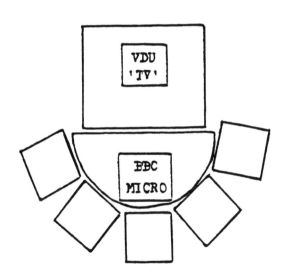

Figure 1b: Semicircular arrangement with two tables and extended monitor
               lead. This arrangement was preferred by the children, and led
               to improved discussion

duals within groups, and for groups themselves, the optimum size turned out to be 4 ($p<0.05$). Individuals in groups of 3 did not work quite as well as those in groups of 4, and in groups of 5 the amount of non-task discussion was greater still. Potter offers the view that in a group of 5 the children at each end of the semicircle tend to become less involved and drop out of the discussion.

In order to prevent this happening, he offers two solutions: first, he suggests that it may be preferable to work in groups of four, because then there is less chance of children becoming isolated in discussion; second, if groups of five are used, the computer should prompt changes of seating, by inviting each child in turn to use the keyboard. Furthermore, these changes in seating should be systematic, with each child moving one place along each time, thus ensuring that no child is away from the centre of attention for very long.

Finally, Potter sought to examine a rather different aspect of children's interaction with computers, namely the extent to which the children saw the computer as infallible, and whether their opinion changed during the time of the project. The fallibility of computers, or more properly, of those who build and program them, was not directly taught, but in certain respects it was implicit in the project. For example, part of one task involved generating new answers, and criticising those of other groups, and those of the teacher, which included one or two answers which were purposely incorrect.

When asked 'Is the computer always right?', the children nearly always said 'No', and tended to give one of three types of explanation. Some were unable to give any clear reason as to why they held this view, and simply rephrased or repeated their answer. Others said that the computer was sometimes wrong, but attributed the failure to hardware problems (for example, one child said: ' ... it might not have been made properly ... the manufacturer forgot something ... the computer might get muddled up ...'). Others gave the answer Potter was looking for, that is, they demonstrated an awareness that if a computer made an error, then a person, probably either the teacher or a programmer was responsible ('[it's] probably not the computer, ... [it's] the people who put the program in ...').

Potter reported that by the end of the project these 9- and 10-year old children were much more likely ($p<0.05$) to give an answer which indicated that they were aware that errors in the computer were probably the result of human errors. He ascribed this improvement not just to exposure to computers, but also to the fact that the children had been creating their own data files, and criticising those of others, so they might have become much more aware that information in a computer was put in by people, and that it was easy to make errors in the process.

1.3.3.2   *Other studies of interaction*

A variety of other studies have been reported in the language and reading field, many of which have been published in the Micro-explorations series (Potter & Wray, 1984; Wray & Potter, 1986).

Trushell and Broderick (1984) report the results of their study of children using word processing software in primary schools in the London area, and indicate a strong preference for collaborative work in writing. Indeed, in a pilot study they had found that children who used word processors as individuals did not make effective use of them. In the pilot study school, access to the word processor was the reward for a good white-paper draft, and as a result, only the better writers were able to use it. Furthermore, the children tended to write poorer second drafts on the word processor, and it took them twice as long to complete them compared with pencil and paper.

In their main study, therefore, children were encouraged to use the word processing program in groups, and to write assignments which were part of their normal class work on a project. They were encouraged to draft on the computer from the outset, and to redraft using a printed version to help them if they wished. Discussion of drafts was encouraged, and the children were taught that everyone had views which were valued, and which would be helpful to the author during the drafting process.

Under these conditions, the word processor was used much more successfully. Over a period of a term, the children gradually came to use the word processor for inserting new ideas at any appropriate point, rather than for simply adding new material at the end. They became much more willing to revise, and made more changes. They also became more willing to postpone giving close attention to copy editing and technical accuracy, and concentrated on more global issues until it was time to produce a final fair copy.

Trushell and Broderick concluded, therefore, that collaborative word processing was likely to be especially helpful for novices, and for inexperienced writers. They noted that under these conditions, children were mutually supportive and co-operative, and far from being intimidated by the new technology, were inquisitive and enthusiastic users of it.

A number of teachers and researchers have considered the potential of computer adventure games in the classroom. Harrison (1984) reported that they led to inventive creative writing at secondary level, although slower learners tended to be conservative in their problem solving behaviour and tentative in drawing maps or diagrams to support their thinking, when using an adult version of a mainframe computer adventure.

Adventure games specially written for children have achieved more success, and publishers have produced a number which not only specifically encourage group work, but focus on specific content and study skills, for example ancient Egypt or marine archeology. Earnshaw (1984) and Johnson (1984) describe the use of an adventure game focussed around a simulated space exploration. They note that the children worked in groups, spending much more time away from the computer than at it, but that the computer provided a highly motivating initial stimulus and an effective aid in rekindling or maintaining motivation during the project.

Johnson's project was in secondary school, and his class made interesting use of two other programs, one of which is a teletext new simulator which offers a similar format to the UK national television teletext systems. The second program was FACTFILE, a database program. In this study, therefore, the children were using one programme to initiate a simulated space exploration, another to write newspaper stories about their adventures, and a third to store and classify information about the fictitious planet they were visiting.

Johnson reports that the project was certainly successful in terms of the systematic use it encouraged of the microcomputer, but it was also valid on other educational grounds: the children's collaboration extended beyond the computer, and groups followed up with library-based research work into aspects of space, computers and robotics.

Bleach (1984) also used an adventure game, and monitored the discourse of 6- to 8-year old children over a period of fifteen weeks. She found that group interaction with the computer proved to be a very valuable social activity, coming as it did at the time when children were just beginning to establish the behaviours of negotiation and turn-taking in personal and social relationships.

The early tape recordings contained a good deal of uninhibited and ineffectual discussion, but over the first six weeks of the project, the computer, and the need for finding a collaborative approach to solving the problems in a simple adventure game, produced a number of changes.

The children became better at turn-taking, and more willing to listen to others, as they realised that valuable information, which might help them solve the puzzles, was being lost if everyone talked at once. Towards the end of fifteen weeks, children became willing to listen to the opinions of those of the opposite sex, which had not been the case initially. As the children became more willing to listen to each other, the mean length of utterances increased, and this also permitted some children to formulate hypotheses which could subsequently be discussed and tested.

Overall, therefore, Bleach argued that the computer could function as an important catalyst in developing a number of social and language skills. Young children, she found, experienced very few problems of alienation or anxiety in becoming familiar with using a computer.

In the UK, teachers of children with learning difficulties are developing a similarly positive view of the microcomputer. Smith (1986) argues that the use of word processing software with slower learners can be very beneficial, and Woodall (1986) describes the use of programs whose primary aim is to encourage oral work among very slow learners, relating particularly to description and ordering of events.

Goulding and Harrison (1987) report a study of the use of a simple adventure game with pairs of slow-learning secondary-age children. Their conclusions were that slow-learning children showed remarkable tenacity and patience in playing the game. During the 430 minutes of tape-recorded activity, there was an unusually high level of attention, and no incidence of off-task activity was recorded. This is remarkable in children who are all taken out of normal schooling because of personality or learning problems. The slow learners were tentative, and avoided much of the exploratory behaviour which a control group of average children demonstrated. Nevertheless, Goulding and Harrison argue that there were essentially few qualitative differences between the two groups; the main problems which the slow learners had were poor processing speed and poor recall. They suggest that similar activities would be beneficial to slower learners, especially since they foster co-operation, and offer a basis for problem-solving in a situation in which risks may be taken safely, and without fear of humiliation or embarrassment.

1.3.4 CODA: WHAT IS THE NATURE OF THE COMPUTER'S INTERACTION?

In the main section of the paper, it has been argued that the concept of interaction is a pervasive and useful one in approaching a consideration of the use of microcomputers in primary and secondary schooling in the UK. Clearly, the argument which has been advanced has taken it as axiomatic that interaction should not be simply between an individual and a computer, but should be much more dynamic, involving a four-way exchange between learner, peers, teacher and computer.

Nevertheless, it seems worthwhile to consider the nature of the contribution to these manifold interactions which the computer can make. My argument is that we should regard the potential nature of the computer's contribution to the interaction as problematic. In other words, I would suggest that, just as it can be facile to describe a computer program as 'good' or 'bad', it would be equally facile to deny that a child's interacting with a computer is not intrinsically 'good' or 'bad' in educational terms, but can be either, depending not simply on the content of the program, but on how that content is mediated.

The 'interaction continuum' which I offer below could serve as a basis for discussing this aspect of the problem. The idea behind it is certainly not new, but I feel it is important, and that it might possibly suggest some areas for discussion which relate to the future development of both programs and computer usage in schools.

The continuum covers four regions:

$\leftarrow$ intolerant $\leftarrow$ tolerant - friendly $\rightarrow$ intelligent $\rightarrow$

We are all familiar with intolerant software. In particular, one can think of programs which are intolerant of case variations (for example, not accepting 'yes' for 'YES'), or which are positively abusive to those who fail ('You fool! You've got yourself killed!! Score - 0 points'). Intolerant software would not accept incorrect spelling or alternate answers.

Tolerant software would be more accepting of case variation, and would be willing to accept spelling errors and alternate answers.

Friendly software would go further, and might offer hints, and correct a person's spelling, It would suggest alternative routes if a person were stuck, and would congratulate students on their success. In short, it would include help systems, and would offer positive reinforcement.

Intelligent software would go further, and would make use of online information, monitoring the user's behaviour and initiating the use of help systems to prompt a better use of the system. It might, for example, construct an online 'process history' of the user's behaviour and compare this with stored templates from other users, and respond accordingly, for example by setting questions to invite a deeper response, by restructuring the task, or by offering text at a simpler level.

Programming costs money, and there is in general a direct relationship between the degree of 'friendliness' or 'intelligence' in a program and the amount of money needed to produce it. Nevertheless, I would suggest that in considering the nature of the interaction between students and computers, we should give serious attention to this issue. In nearly all the studies reported in this paper, the computer takes on some aspects of the role of a teacher, and it is crucial, therefore, to consider whether that teacherly aspect of the computer's role is morally and pedagogically defensible. This is an aspect of interaction which we ignore at our, and our students', peril.

1.3.5   BIBLIOPGRAPHY

Adams, A. and Jones, E., Teaching Humanities in the Electronic Age, Milton Keynes: Open University (1983).

Bleach, P., Using Magic Adventure in the Classroom. In Potter and Wray (op. cit.) (1984).

Burkhardt, H. (ed.), Design and Development of Programs as Teaching Material, London: CET (1982).

Earnshaw, P., Using an adventure program. In Potter and Wray (op. cit.) (1984).

Goulding, S. and Harrison, C., Problem solving: slow learners using a microcomputer adventure game. In B. Smith (ed)., Microexplorations (3), Ormskirk: United Kingdom Reading Association (1987).

Harrison, C., Reading and the microcomputer: more answers than questions? Paper given at IRA convention, Atlanta, GA (1984). In B. Smith (ed.), Microexplorations (3), Ormskirk: United Kingdom Reading Association (1987).

Harrison, C., Criteria for evaluating microcomputer software for reading development: observations based on three British case studies. Journal of Educational Computing Research 1 2, 221-234 (1985).

Hope, M., Article in The Times Educational Supplement, 13th September (1985).

Johnson, B., Spacex: using the adventure game in the secondary classroom. In Potter and Wray (op. cit.) (1984).

National Association for the Teaching of English, English Teaching and the New Technology: Into the 1990s, London: Council for Educational Technolgy (1986).

Potter, F., Classroom organisation and group discussion: the role of the microcomputer, the role of the teacher. Project Report - Edge Hill College of Education (1985).

Potter, F. and Wray, D., Microexplorations (1), Ormskirk: United Kingdom Reading Association (1984).

Smith, B., The use of the word processor with developing writers- helping children to think about writing. In Wray and Potter (op. cit.) (1986).

Stake, R.E., The countenance of educational evaluation. Teachers College Record, 68, 532-540 (1967).

Taylor, J., Using a computer for reading and language development. D.A.S.E. dissertation, Edge Hill College of Higher Education (1985).

Trushell, J. and Broderick, C., Primary observations of word processing. In Potter and Wray (op. cit.) (1984).

Woodhall, K., The computer as a springboard for developing oral language skills. In Wray and Potter (op. cit.) (1986).

Wray, D. and Potter, F., Microexplorations (2), Ormskirk: United Kingdom Reading Association (1986).

1.4 SWEDISH IMPLEMENTATION OF COMPUTER APPLICATIONS INTO THE EDUCATION OF THE NINE-YEAR COMPULSORY SCHOOL

by

Dr. GÖRAN NYDAHL,

Ministry of Education, Stockholm

1.4.1   SUMMARY

The subject of computer techniques in schools already began to attract interest in the 1960s, but it was not until 1971 that the Government instructed the NBE to investigate the educational implications of computers in schools. This led to the DIS-project, the results of which were presented in 1980. The DIS-group recommended that pupils in grade 9 should be given basic instruction, corresponding to 20 teaching periods, on computers and their use in society, i.e. computer science.

In 1983 the Government and Riksdag raised their sights by augmenting computer science instruction to 80 teaching periods in grades 7 - 9. To support the implementation of this scheme, a special syllabus was drawn up, defining the content of instruction, and also including recommendations on teaching strategies. Municipal authorities were awarded State grants towards the cost of hardware, and supportive material (suggestopedias, etc.) was compiled. A special training scheme, corresponding to ten training weeks, was introduced for teachers who were to provide their colleagues with support and assistance.

The Government and Riksdag are sceptical about using computers in grades 1 - 6. First, they want to see more research and development concerning the effects of computers on children in these age groups. The economic consequences of extending CAT to these grades are another reason for hesitancy.

Computerised instruction in computer science involves using ready-made instrumental programmes, such as word processing programmes, spread-sheet programmes and database programmes. These have proved to be viable forms of computer support for various subjects. Computer science teaching has gradually resulted in a growth of interest in the use of computers in other context besides computer science, thus generating a need and a desire for more interesting computer applications in schools.

The Educational Software Group, set up at the Ministry of Education in the autumn of 1985, is conducting a number of projects and development schemes aimed at developing good pedagogics for the computer as an aid to teaching. Special interest is also being devoted to new types of programmes and media.

The Government is drafting a Bill for the expansion of experimental computer support in all grades of compulsory school over a five-year period. If the Bill is passed by the Riksdag, the enlarged experimental scheme can come into operation in the 1988/89 school-year.

1.4.2   INTRODUCTION

Sweden has one of the world's highest computer-population ratios. Swedish enterprise bases its competitive strength on a high level of technology, and on a wide distribution of skills in its workforce. This basic view is shared by politicians, added to which the Government and Riksdag (Parliament) believe that the system of democratic government can only be preserved and developed if citizens acquire the knowledge and ability to make use of the new information technology.

In keeping with this fundamental conviction, an ambitious project has been started to give both young persons and adults an education adapted to the emergent computer and information society.

Quite early on, this programme acquired distinctive characteristics, and it has sometimes been dubbed "the Swedish model". It is intended to apply to all aspects of information processing, but it is particularly in evidence in the educational sector. In a word, it implies taking the following points into account, in one single context, where information technology questions are concerned:

-    the technology itself;
-    its utilisation;
-    the effects of its utilisation on the individual, the organisation and society.

1.4.3   DEVELOPMENTS UNTIL 1983

A number of educationalists, both in individual schools and the national school administration, realised already in the 1960s that computer technology was bound to have an impact on teaching.

In 1971 the NBE was instructed by the Government to investigate the faisibility of teaching computer technology etc. in compulsory and upper secondary schools, and the DIS-project (DIS being the Swedish acronym for The Computer in School) was started in 1974 to investigate the educational implications of computers in schools, i.e. the effects of computers on the content, method and organisation of teaching and on in-service training and teaching materials.

DIS presented its findings in 1980, and they form the basis of the NBE infotech action programme, which can be briefly summarised as follows:

-    All pupils in grade 9 of compulsory school were to be given basic instruction concerning computers and their use in society - computer science. Computer science was to be incorporated into existing school subjects.

29

-    The programme recommended supportive measures with regard to hardware, software, INSET and basic teacher training.

The DIS-group's recommendations were already assured of political support at the time of their publication. Although the recommendations led to the introduction of compulsory computer science in Swedish schools, ambitions were in several respects limited. For example, it was not considered necessary for compulsory schools to acquire their own computer equipment and, volumetrically, compulsory instruction was expected to occupy about 20 hours at the senior level of compulsory school. Individual pupils, however, could obtain a great deal more instruction than this by means of "free" activities and elective studies.

The early 1980s brought the presentation of various central research and development projects which were to leave their mark on subsequent developments.

The PRODIS-project (Software and hardware in schools) was mounted by the NBE to specify functional requirements for hardware and software in upper secondary schools. Requirements were defined for different subject fields, for the guidance of municipal authorities in their purchasing of equipment.

The PRINCESS-project (Project for Research on Interactive Computer-based Education Systems), in progress since the early 1970s at the Department of Automatic Data Processing, Stockholm University, ended in 1983 with the presentation of a report entitled 'Computer Support in Education'. That report contains an emphatic warning of the danger of information technology leading to a rebirth, in new guise, of the old teaching machines. Computer support should be used to give pupils an opportunity to solving problems in teaching; it must be a supplement to teaching, designed in such a way that the user is at liberty to use it and modify it. The Government and Riksdag have endorsed this view, making it an integral part of their policy on the use of computers in education.

The TUDIS-project was a technology procurement project, conducted by the Board for Technical Development (STU). This led to the prototyping of a school computer, Compis.

Increasing amounts of computer instruction, both in traditional disciplines and in new subjects and courses, generated a heavy demand for in-service training. Teachers' infotech courses, varying in duration from 1-20 weeks, were given priority from the outset and accounted for roughly a quarter of all INSET weeks at higher education establishments. Local courses were also organised.

1.4.4   RISING AMBITIONS - DEVELOPMENTS SINCE 1983

In 1983/84, the Government and Riksdag raised their sights higher than ever, prompted by the rapid transformation of working life, the availability of powerful personal computers at a price making them affordable to a larger number of schools, including even compulsory schools, and a declaration of political intent by the new Social Democratic Government returned in 1982.

30

The NBE organisation was reinforced in response to these new signals. A special programme, Education for the Computer Society, was launched in 1983, and that same year a policy document was adopted.

The most conspicuous manifestation of the new aspirations was the development of computer instruction at the senior level of compulsory schools resolved on by the Riksdag in the spring of 1984. That policy resolution provides for the computer science teaching time allocation at senior level (grades 7 - 9) to be increased to about 80 hours over a three-year period, starting in the 1986/87 school-year. The following were among the measures taken to support the introduction of computer science:

-       A special plan of studies was compiled, closely defining the content of instruction and offering practical recommendations.

-       State grants were awarded to municipalities purchasing computer equipment. To qualify for grants, this equipment has to meet certain minimum requirements as regards both computer system and hardware.

-       Special ten-week facilitator training schemes were started to give schools the necessary teaching competence. The intention was for one or two persons from every senior-level school to complete these training courses. A certain amount of previous knowledge was required.

-       Central supportive material was produced in the form of suggestopaedias and method books.

1.4.5   PROBLEMS AND QUESTIONS OF PRINCIPLE

The introduction of computer science in school raises many questions and quite a few problems. Here are some of the points which have been observed in the Swedish debates, together with the preliminary standpoints adopted by thee Government, the Riksdag and central authorities.

1.4.5.1   *What ought basic computer education to include, so as to prepare students for the computer society?*

The answer arrived at in the Swedish debate is stated in the DIS-report from 1980, the compulsory school computer science syllabus from 1984, and the official report on broad-based computer education from 1985 (Bred datautbildning SOU, 1985:50). All these documents emphasise the democratic element.

The report on broad-based computer education, finally, contrasts two cognitive perspectives which are said to occur in the existing educational material. One of these points to the possibility of individuals and groups of persons influencing the design of computer systems. In addition, facts are problematised and related to a wider social, economic and political context. The other perspective is more depictive, emanating from current conditions and trends. The official standpoint is that educational content must be designed so as to stress the first of these perspectives.

31

1.4.5.2 What part is to be played by programming, and which programming language or languages are to be chosen?

The compulsory school syllabus tones down the importance of programming, while putting more stress on the use of standard software, a general knowledge of information technology, and consideration of social aspects. Actual developments in schools are moving in the same direction.

The launch of the Compis computer was accompanied by the introduction of a new and much-debated programming language, Comal. Although the new language was more capable of giving programmes a good structure, the critics disapproved of introducing in school a language not widely used outside school. The diminishing importance of programming has made the question of programming languages less interesting. Basic and Comal co-exist, and some schools are also using Pascal.

1.4.5.3 In what grade is computer education to begin?

Computer education now begins, as a rule, in grade 7. The syllabi require computer science to be taught at senior level, i.e. in grades 7 - 9, but schools decide locally the grades in which instruction is to be provided. Some schools do not start until grade 8 or 9. By the time the three-year venture has been completed, there are not likely to be many schools deferring computer science until grade 9.

Many voices are now being raised in favour of introducing computer education at junior and intermediate levels (grade 1 - 6). Reference is made, for example, to the receptiveness of pupils at this age, and to the possibility of counteracting sexual bias in recruitment for infotec jobs.

A certain amount of experimentation has begun at the Department of Automatic Data Processing in Stockholm, and also on a joint basis by the Stockholm Institute of Education, Falun/Borlänge University College and Linköping University.

The official policy, so far, has been one of "wait and see". Efforts at the senior level of compulsory school demand heavy resources and must be given priority. Besides, we know very little about the possible effects of small children on working at computers, and it is uncertain what instruction at junior and intermediate levels would include. This makes it essential for research and development to be undertaken by many different parties before any more far-reaching decisions are made. Experimental and more developmental activities will be expanded in the next few years.

1.4.5.4 What hardware is suitable for schools? How is this aspect to be controlled?

The official line in Sweden has been that heavy demands must be made on equipment, even at compulsory school level. This has been reflected, for example, by the specification of standard requirements, and the restriction of State grants to equipment tested and found to meet those requirements.

An initial specification of requirements for compulsory school was adopted in 1984. So far, about 15 makes of computer have come to be included in the list of brands qualifying for State grants.

Unlike some other countries, then, Sweden has decided against severe top-down restrictions on the number of computer brands to be used in schools. The Compis computer developed with State support is having to compete with other brands in the school market. There has been some standardisation, however, above all through the stipulations made concerning operating systems (at least CP/M or MS-DOS or a compatible type).

The practice of premitting several different operating systems involves manifold disadvantages. Because there are so many suppliers, each supplier achieves less volume and profitability on school projects. INSET, which is procured externally, is more liable to be based on equipment differing from that which the teachers have in their own schools. Co-operation between schools is impeded.

More weight, however, has been attached to the benefits. These include the competition maintained between producers, which will also benefit schools in the long run. Then again, it is natural that computer installations in schools should reflect the computer mix in the employment sector. And in many cases the expense of developing software for one single computer to be used only in schools would be prohibitive.

A new specification of compulsory school requirements was drawn up in the autumn of 1986. This stipulates, for example, that the computer system must be able to work with MS-DOS and must be capable of handling a specified graphics standard. Today this does not appreciably limit the number of computer makes, while on the other hand, it provides a foundation for the uniform development of software for schools.

1.4.5.5   What software is suitable for schools? How is it to be procured?

One general question here, is whether one should invest in special software for schools or whether schools should try to use the same software as in the employment sector.

One is bound to answer both yes and no. Schools must prepare students for working life, but the principles of the school curricula require schools to make special demands where software is concerned. Standard software may be found to fall short of school requirements. Exacting requirements in this respect can have a positive impact on the general standard of software in the community at large.

But how is the software to materialise? Hopes of this problem being solved by market mechanisms have not been borne out by events. Public measures are needed. An initial step was taken in 1986 with the formation of the Educational Software Group.

The Educational Software Group is investigating fields in which schools require software, and the demands which the software will have to meet. In addition, it is engaged in the inventory and testing of suitable software for schools. Its remit also includes the initiation of research and educational development work to augment our knowledge of the potential educational uses of the computer.

1.4.6 COMPUTER-ASSISTED TEACHING

The development of computer education in compulsory school mainly involves the compulsory school subject known as computer science. In principle, therefore, computer-assisted teaching has to make do with the computer time left over from computer science lessons, and is thrown back on local initiatives and resources.

Computer science includes word processing. The natural means of imparting substance to this instruction is by using examples which support the pupils' written work in Swedish, as well as getting the pupils to use computers in compiling reports on a social or scientific subject. Used in this way, word processing makes a powerful form of computer support for teaching.

An equally natural link should exist between calculation work and the teaching of mathematics, civics, etc. Work on databasis can very well be used as a means of facilitating modern pedagogics based on current information in certain social subjects. Control and measuring techniques can be used for interesting experiments in the teaching of physics, for example.

It is natural that these developments should make teachers more interested in computer support in other teaching contexts as well. But everybody is anxious to maintain a high standard of objectives for different forms of computer support in schools. These objectives are usually formulated as follows:

The computer should only be used as a teaching support when it is calculated, one way or another, to expand the scope of teaching or to improve its quality, and computer programmes must be constructed to as to avoid imposing narrow constraints on the pupil.

This makes instrumental programmes the most important type where schools are concerned. Directly instructional programmes of the drill and practice type normally have little appeal for schools in Sweden.

In keeping with this approach, the Educational Software Group at the Ministry of Education is conducting a large number of development projects at various schools throughout the country. The problem often lies, not in the development of new software, but in the compilation of educational experiences surrounding programmes already existing in schools. This experience will be successively documents in a number of reports for the support and assistance of teachers and teacher trainers.

1.4.6.1 *Examples of development and research projects initiated and/or supported by the Educational Software Group*

Grades 1 - 6

Pending new research findings and an impending declaration of political intent, only a few projects have been started in these grades.

Stenhagsskolan in Akalla (grades 1-3) and Billingskolan in Skövde (grades 1-3) are working mainly on word processing as a means of improving and stimulating pupils' reading and written work. Briefly, these projects are based on pupils writing for each other, not just for the teacher. Importance is attached to classmates reading and commenting on each other's work.

At Hagenskolan, near Göteborg (grades 1 - 3) LOGO and specially selected training programmes are being used in class.

A number of Jönköping schools (grades 4 - 6) are using videotex databases for teaching purposes. Teaching is being made to concentrate, among other things, on new information in the database, and on a communication system for transmitting simple messages between different school classes in Sweden. Within the database, the participants are also building up a database of their own, containing information about different countries in the world. This project has been made to focus primarily on Africa. The database, based on simple frame images (such as Prestel in the U.K.), is inexpensive to use and does not require any costly hardware.

Grades 7 - 9

Most schools with these grades have modern computers and a number of teachers with D.P. training. Naturally, therefore, most projects are located in these schools. Most of the programmes used are open ones of the instrumental type.

-    A number of programmes are being tested in mathematics teaching. One of them, called "Mathematics Workshop", is particularly interesting. This programme covers practically the entire theory of functions. With it the pupils can draw arbitrary functions, calculate functional values, differentiate, integrate, and so on. The educational approach is for the software to enable pupils to concentrate on the mathematical problem. Once a pupil has understood and formulated the problem, the software is used in support of the calculations that follow. This educational approach is a topic of lively debate among maths teachers.

-    In a project directed by the Educational Software Group, a study is being made of various ways of using calculating programmes in teaching. Teachers are developing new examples to provide colleagues with a suggestions bank. This project has also resulted in the observation that use of complete or semi-complete calculations can often have an instructional value.

-    In Swedish teaching, of course, a study is being made of work processing as a potential support for the development of writing skills.

-    In one project, focussing on modern languages teaching, an attempt is being made to develop a programme idea in support of a communicative approach to FLT (Foreign Language Teaching). The aim is for pupils to experience the necessity of adapting their dialogue to the current situation and to the person they are addressing. What happens if you "get it wrong", how does an adult react when spoken to like a classmate, and so on?

-    One big project concerns the use of databases for civics teaching. Here the educational approach is concerned with placing information relevant to problems normally discussed in civics lessons, but from a central school database. Typical problems: "What is the reason for world starvation?",

"How do living conditions in this community compare with other parts of Sweden". Once the information has been located in the database, it is taken down to the school's local computer system, where it is processed further before being presented, together with other material, in the school report on the problem.

-    A number of simulation programmes and hardware for control and measurement techniques are being tested in science teaching at a number of schools.

-    In a project at Umeå University College, a study is being made of computer support as a means of diversifying art education. This is based on the observation that computer images today belong to the pictorial language of the society in which we live. It is important to know how these images are produced and how they can be used and exploited in the community. Certain computer applications also give pupils in schools an opportunity of creating and elaborating images of their own, and these too are being studies in the course of the project.

-    At Maserskolan, Borlänge, a comparison is being made of the development of music classes receiving computer-assisted instruction, and classes taught by more conventional methods. The computer-assisted teaching items are concerned with "creating music", "musical form", and "making music together".

1.4.6.2   *Projects of general interest*

Program design

In Sweden, just as in the other Nordic countries, there are special programme design courses. These are based on a method developed partly by Les Green, Ontario, Canada. The method begins with an analysis of the computer support which teaching requires. At attempt is then made to develop some of these requirements into a programme idea by blocking out a comprehensive picture of the workings of the programme. Development proceeds by various phases, until a concept is evolved which is ready for programming. Programmers are not called in - if at all - until the programme idea is fully developed.

This design method leads to the development of a number of programmes, but many people would say that it is the actual design process that really counts, because it gives teachers a more advanced appreciation of the use of computer programmes in teaching.

Expert systems in teaching

In another project, headed by the Educational Software Group, a study is being made of potential uses of expert systems or knowledge-based systems in teaching. This is a two-pronged project. One aim is for the pupils to construct systems of their own, while the other is for ready-made systems to be introduced as supplementary teaching materials. As part of the project, the following hypotheses are being investigated:

-    If pupils are given the task of constructing expert systems, they have to begin by acquiring the relevant knowledge and then structuring it carefully into rules and facts. This approach compels the pupil to

36

understand the field of knowledge concerned. The computer will not accept any gaps in the pupil's knowledge.

- The pupils' acquisitions of knowledge is a continuous process in school. A cognitive process is built up successively and a small base can be augmented and improved. This characteristic of the shell of an expert system provides a natural form of support for the pupils' learning.

- Advanced expert systems can be used in teaching to make new knowledge available and present it in a new fashion. In this way, the system can provide an important complement to other teaching materials.

Interactive video disks in teaching

Advanced techniques for using laser video disks have existed for a couple of years now, but It has proved difficult to devise really good teaching applications. The first clear indication of potentially interesting applications for compulsory school teaching came with the BBC Domesday Project in the U.K.

Inspired by the Domesday Project, the Educational Software Group has started an attempt to produce a video disk describing the Old Town district of Stockholm, past and present. Pupils will be able, for example, to go on a simulated sightseeing tour of a number of streets in the Old Town by advancing through a sequence of images. If they want to study a house, all they have to do, for example, is to point a 'mouse' at the front door. The house can be investigated as it is today, and as it looked and was used at different times in the past. The disk will make it possible for teachers and pupils to plan their own picture sequence for sightseeing, which can then be used to illustrate a teaching point, or as an illustrative companion piece to a talk given in class.

The disk will also describe the lives of two people living in the Old Town during the 1850s, and it will illustrate their working and living conditions. The two people concerned have been selected from the Stockholm Historical Database, the criteria being that one must be a man, the other a woman, and both must have spent their entire lives in the Old Town.

The most time-consuming part of the project soon proved to be the collection and cataloguing of the illustrations, added to which, picture descriptions and overlay texts take a very long time to produce. The aim is for the disk to contain between five and ten thousand still pictures, and about 20 minutes of moving sequences, divided between the same number of events of descriptions.

1.4.6.3   *Viewpoints on computer support in schools*

The use of computers in teaching is a moot point in schools. Enthusiastic teachers taking part in projects come up against scepticism and apprehension on the part of other colleagues faced with the new technology. "What good does it do, using computer support in a particular situation?" "Teaching is getting far too dependent on high technology." "How does computer support affect the pupils' basic skills?" "How can teachers find the time and the energy to assimilate every newfangled idea?" These are some of the questions and viewpoints emerging from the debate. Excessive confidence in technology on the part of some is offset by a natural scepticism on the part of others.

Many computer applications in school have to be crammed in on top of traditional teaching. The result is shortage of time and, often, bad pedagogics. Teachers must have the courage to substitute new teaching for old. It follows that a completely new pedagogics must be used, because the computer is often incapable of copying old teaching methods.

New pedagogics calls for INSET. In the future, specialised infotech courses and nothing else will not be the right response. Instead, things are moving in favour of computer application being made a natural, integral part of normal subject-related INSET. A dialogue has been inaugurated between the Educational Software Group, the National Board of Education, and those responsible within the higher education system, to discuss a new emphasis in future INSET courses for teachers.

One conclusion on which most people appear to agree, is that short learning programmes, often referred to as "lesson ware", are of limited usefulness in schools. The value of computer support lies in different types of instrumental programmes, facilitating new pedagogics.

1.4.7   LOOKING AHEAD

All sectors of society are feeling the effects of a more adverse economic climate. In schools there is sometimes talk of amenities being dismantled. Funding for expensive computer equipment is not always readily available. Sometimes the debate in schools results in a handful of teachers who are interested in computers, finding themselves at loggerheads with a majority of their colleagues, who feel that their own teaching material requirements are being neglected.

As computer support comes to be more widely used in teaching, schools will need access to more hardware and software. Instead of just being installed in computer rooms, as at present, computers will also have to be readily available in every classroom. Heavier investment in equipment will have to be based on a policy decision, probably in statutory form.

The highly preliminary findings obtained by the Educational Software Group show that computers can be used by many teachers of different subjects, as a means of improving the quality of teaching. By extending the use of computers to more teachers, and by augmenting the use of computers in preparatory work, hardware and software can be made interesting for all teaching purposes and, accordingly, attractive to all teachers.

A report, Action Programme for Computer Education in Schools, Adult Education and Teacher Education (DsU 1986:10), presented to the Government in the autumn of 1986, recommends the allocation of funds over a five-year period for an intensification of CAT (Computer-Assisted Teaching) experiments in all grades of compulsory school. The teachers involved are to be given special in-service training, and funds are to be made available for hardware and software. The National Board of Education is to take over from the Educational Software Group, and is to allocate money for directing and conducting this intensified experimentation. The Government will be introducing legislation on the subject, probably in the autumn of 1987. Given Riksdag approval, experimental activities can be started in the 1988/89 school-year.

Only after several years' experimentation can the Government and Riksdag be seriously expected, if at all, to adopt a standpoint on the more systematic use of computer-assisted teaching in Sweden.

1.5 ASPECTS OF THE INTRODUCTION OF NEW TECHNOLOGY INTO EDUCATION IN FRANCE:

INTERACTIVE LEARNING AND LEARNING INTERACTION, THROUGH RESEARCH CARRIED OUT AT THE NATIONAL INSTITUTE FOR EDUCATIONAL RESEARCH

by
Dr. BERNARD DUMONT,
National Institute for Educational Research,
New Technology and Education Programme, Montrouge, France

1.5.1   SUMMARY

The purpose of the New Technology Education Department of the National Institute for Educational Research (INRP) is to study the impact of communication and information technology on the French education system, by means of research carried out in conjunction with teachers and pupils at various educational levels.

Research workers from this Department help not only to evaluate the introduction of new technology in educational establishments, and to devise and test educational products and field experiments, but also contribute to working out concrete proprosals for both policy makers and teachers, and take part in training programmes for teaching staff.

We present the research, completed or ongoing, in the following three fields:

- Microcomputers (physical sciences, natural sciences, remote sensing, mathematics, humanities);

- Information technology (video-communication, viewdata);

- Audiovisual technology (television, video disc);

- New research (artificial intelligence, IT-networks, technological imaging).

We attempt to show how interactive learning has been implemented, using new technological tools, and to demonstrate the need for interactive learning for both teachers and pupils.

To carry out the research objectives of the Department, 27 full-time research workers, 5 part-time teachers/research workers, and almost three hundred associated teachers are studying different examples of the introduction of new technology in schools.

From schools television (c.f. the "Jeune Téléspectateur Actif" experiment) to the introduction of video discs and "minitels", through the various computer schemes of the Ministry of Education, the INRP researchers help not only to evaluate experiments, and to devise and test educational materials, and field experiments, but also contribute to working out concrete proposals for both policy makers and teachers, and - where applicable - take part in training programmes for teaching staff.

1.5.2   INTRODUCTION

We would now like to outline the completed or ongoing research in the following three fields: microcomputers, information technology, and audiovisual, and we would then like to indicate the direction in which new research is heading.

The following examples attempt to illustrate what we mean by "interaction";
this does not mean allowing the learner to type "yes" or "no" or a number
from 1 to 4 on the computer keyboard, but putting the learner in an active
decision-taking situation, a "research"-situation: acquiring knowledge,
exploring different environments, simulating experiences, and processing
information.

This explains, on the one hand, the almost total absence of tutorial software developed or studied in our Department and, on the other hand, why
we are concerned with training tutors, with a view to collective integration
of these new tools and developing a real interaction in classes.

1.5.3   MICROCOMPUTERS

The Computers for All-Scheme ("Plan Informatique pour Tous"), started in
1985, enabled all schools to present pupils, in both primary and secondary
education, access to microcomputers.

The scheme also enabled thousands of teachers to gain a basic training in
using computers for educational purposes.

The INRP wanted to go beyond the computer-assisted teaching aspect, by
carrying out research where the computer would not merely act as a tutor
but would try, on the one hand, to offer original solutions to standard
educational problems and, on the other hand, to play a role in the emerg-
ence of new methods in learning new concepts or new skills.

1.5.3.1   *The computer in science laboratories: a data acquisition and cal-*
culation tool

This research involves exploring the possibilities of using microcomputers in
physics teaching, based on:

(a) The omnipresence of computing in professional laboratories;

(b) The increasing diversity of places where computers are installed, as
the need for computers becomes more diverse. Various types of software and hardware have been developed and put to the test in classes, especially in mechanics, e.g.:

-    explosions on an air-cushion bench (introduction to the notion of
momentum);

- shocks on an air-cushion bench (conservation of momentum);

- free-fall of a ball-bearing (the law of motion, conservation and energy).

Similar work has been carried out at the "Conservatoire National des Arts et Métiers" (College of technology for training students in the application of science to industry; see bibliography below).

As well as giving the physics pupil a better idea of what is, at the moment, a physics experiment, a computer allows for better distribution of work during practicals, between carrying out experiments, analysing results and interpreting them.

As regards teachers, a computer is mainly appreciated for its capacity to measure new quantities with great accuracy, and then reproduce them in the form of graphs or statistics, leading to a new approach to the notion of measurement variation in physics.

The interactive aspect is clear in such activities, where the learner is relieved of tedious tasks, but is obliged to use his new tools "intelligently", whence the need for "learning interaction": e.g. how to modify the conditions of an experiment, taking into account results already obtained from previous experiments, how to structure the data collected, how to interpret them statistically and graphically.

Bibliography

F-M BLONDEL, M SCHWOB, Du laboratoire à la salle de cours: utilisation de l'ordinateur en sciences physiques. In: Ordinateurs in physique-chimie ou comment s'en servir dans l'enseignement. INRP-UDP, 1985, pp.9-39.

F-M BLONDEL, J-C LE TOUZE, Acquisitions de données par micro-ordinateur In: idem, pp.58-69.

F-M BLONDEL, S SCHWOB, Etude des utilisations de l'informatique dans l'enseignement de la physique au lycée. In: Revue Française de Pédagogie, No. 72, 1985.

F-M BLONDEL, J-C LE TOUZE, N SALAME, Ordinateur et expérience de mécanique. In: Actes des 2èmes Journées "Informatique et pédagogie des Sciences Physiques". Nancy, 21 and 22 April, 1986, pp.70-75.

F SOURDILLAT, C RELLIER, Ordinateur, outil pédagogique de traitement et de visualisation de mesures et de commande au laboratoire, à l'atelier, dans la classe. In: idem, pp.76-88.

F-M BLONDEL, J-C LE TOUZE, N SALAME, Ordinateur et activité expérimentale en physique. Exemples de mécanique. In: Bulletin EPI, No. 42, June 1986, pp.75-82.

Ordinateur outil pédagogique au laboratoire et à l'atelier. Collective publication, CNAM, Paris, January, 1987 (5èmes Journées Nationales de Synthèses et d'Etude).

1.5.3.2 *The computer and national science: learning and consolidation of the experimental method*

Research carried out in the context of using microcomputers in teaching natural science, in both lower and upper secondary schools, angled towards increasing the possibilities for experiment in the field of complex system dynamics (physiology, ecology).

It has resulted in a dozen software products currently being used in ordinary schools and in teacher training.
Two aspects are covered:

(a) Simulation: to encourage understanding of experimental reasoning, through activities relating to requests for additional information, selecting and testing hypotheses, and choosing exploration methods;

(b) Databanks:
- to be built up by the pupils;
- larger ones published by scientific bodies and research laboratories (for example: the Physical Cemical Data Bank for Lake Geneva, established by the INRA (National Agronomic Research Institute) in Thonon).

These two approaches enable pupils to put into action basic methodological procedures in experimental science: sorting, organisation, relevance of data to be taken into account, various processing of data.

In both cases, the learner is in an "active" position, and cannot merely follow a route preset by the software programmer, because such a route does not exist. The pupil, therefore, has to learn to use the resources placed at his disposal by the microcomputer; his approach has to be thoughtful and not random, or he risks being swamped by unstructured, irrelevant data.

Bibliography

Informatique et enseignement des sciences naturelles. Collective publication, not part of the collection. INRP (Communications présentées aux Journées de Sèvres, 18 et 19 juin 1984).

M DUPONT, N SALAME, Simulation d'expériences au contrôle du raisonnement. Un exemple en endocrinologie. 8èmes Journées Internationales sur l'Education Scientifique, Chamonix, February, 1987.

R CULOS, S DUPOUY, J-P MAURIES, S SABATIER, TRAME: un logiciel pour l'étude de la transformation métamorphique des roches. In: Bulletin de l'EPI, No. 45, March, 1987, pp.113-120.

M ALCHER, I BAVEUX, F-M BLONDEL, N SALAME, Simulation en biologie: aide à la déduction dans le raisonnement expérimental. In: Actes du 2ème Congrès francophone sur l'EAO, Cap d'Agde, March, 1987.

F-M BLONDEL, N SALAME, Simulation and reasoning in biology: towards intelligent learning environments. CAL 87, Glasgow, April, 1987.

1.5.3.3 *Experimental introduction of remote sensing in secondary education*

The main aim of this research is to devise and experiment with ways and means of introducing aerial and spatial remote sensing into secondary education, from a multi-disciplinary point of view, to study whether a new type of data would be useful to pupils and teachers, and to propose possible syllabus changes to take this new technology into account.

Owing to the innovative nature of the field of study and investigation methods, research with regard to pupils could not begin until after a training period for the teachers. At the moment, several experiments are being carried out on different themes concerning physics (especially radiometry), natural science, geography (from, inter alia, its economic, historial and even mathematical aspects): satellites, images and maps (cartography, climatology, metereology), study of a natural forest environment, photosynthesis, agricultural statistics, urban development, hydrology, etc.

Within this framework, analysis and image processing software has been produced and digital images have been transferred to appropriate media. Data have been provided by the National Space Studies Centre ("Centre National d'Etudes Spatiales") (SPOT satellite), and by field studies.

Results of this research have now been published by the INRP, as part of the "Rencontres Pédagogiques" series.

Bibliography

S DUPOUY, La télédétection au lycée. In: Espace Information, Bulletin du
    CNES, No. 31, October, 1985, pp.22-49.

S ESTIVAL, A HIRLIMANN, M VAUZELLE, Télédétection numérique et en-
    seignement secondaire. In: Bulletin de l'EPI, No. 38, June, 1985.

1.5.3.4 *Collective use of the computer in mathematics lessons*

Work has been carried out with teachers/research workers from the Centre for Research and Experiment in Mathematics Teaching of the "Conservatoire national des Arts et Métiers" and the Institute for Research on Mathematical Education of the University of Paris 7. The aim of this research was to create software for collective computer use in mathematics lessons, as opposed to conventional tutorial use. A computer allows the teacher or pupil to construct and manage geometrical figures with specified parameters on several screens situated in the classroom. The software does not ask direct questions, but the situations which appear on the screen can lead pupils to ask questions or submit theories and conjectures themselves, then, where appropriate, to demonstrate the suggested properties.

The role of the teacher-game leader is fundamental in this form of computer use, if learner interaction is to have its place. Tutors must have special training in the use of "tools".

This software - dubbed "imageware" ("imagiciels") - and aimed mainly at lower secondary, but also at upper secondary schools, owes its distribution largely to the Computers for All-Scheme.

Bibliography

Imagiciels-Enseignement des mathématiques illustré par l'ordinateur, collective publication, collection Rencontres Pédagogiques, 1983, 127 p. (INRP).

1.5.3.5 *Microcomputers in humanities teaching*

Various research work has been carried out in this context:

1.5.3.5.1   At primary education level

A software programme called LOGOGRAM has been designed and tested for use by children; it consists of a declaratory environment where "grammars" of varying complexity can be constructed, intended for a sentence generator or analyser. Although written in LOGO, language available on existing microcomputers for primary education, and with which it is relatively easy to work in a "declaratory" fashion, the user does not have to be able to programme in LOGO.

The aim of this software is to give the pupil a tool with which he can produce or recognise sentences on the computer following a syntax model that he will have been able to define himself beforehand. Thus there is a certain stimulation of language production where, it must be emphasised, the child can create the model, by experimenting, as the sentences produced confirm or invalidate the model imagined.

Bibliography

F ROBERT, LOGOGRAM - Un environnement déclaratif en LOGO, pour construction de "grammaires" destinées à un générateur ou un analyseur. Présentation et manuel d'utilisation, December, 1986, INRP. Not part of any collection.

1.5.3.5.2   At secondary education level

Research is currently being carried out in conjunction with the CNRS (National Scientific Research Council - Institut National de la Langue Française), with a view to using computers to study texts in history, literature and philosophy.

The computer, thanks to software designed by the INRP-team, makes it possible to do mainly vocabulary processing automatically on previously recorded texts.
At the same time, various methods of approaching texts in the following three fields are being tested:

- "Les Fleurs du Mal et les Poèmes en Prose" by Baudelaire;
- Human Rights Declaration;
- Texts on the 1914-1918 War.

Bibliography

Utilisations pédagogiques des banques de données. Paris, Associations Enseignement Public et Informatique, 1984, 208 p.

Des textes avec ou sans ordinateur. Collective publication, Collection "Rencontres Pédagogiques", 1983, 120 p.

P MULLER, Deux approches pour l'étude des textes. In: Le Français dans le monde, September, 1985, pp. 45-48.

Informatique et orthographe. Collective publication, Collection "Rencontres Pédagogiques", 1985, 128 p.

P MULLER, A new pedagogical approach to the study of texts with a microcomputer. In: Computers and the Humanities, vol. 20, No. 3, July-September, 1986, pp. 203-206.

P DAUTREY, P MULLER, Etude des structures temporelles dans le discours politique. In: Méthodes quantitatives dans l'étude des textes. Actes du Colloque International CNRS, Université de Nice, 5-8 June, 1985. Slatkine-Champion, Genève/Paris, 1986, pp. 203-206.

P MULLER, Sur les chemins du ciel. In: Le Français aujourd'hui, No. 77, March, 1987, pp. 27-33.

P MULLER, Lectures comparées des Déclarations des Droits de l'Homme. In:
Le Français aujourd'hui, No 77, March, 1987, pp. 35-38.

1.5.3.6. *Study of the role of a "transparent" computer in primary schools*

This research (sponsored by the CNRS) is examining, with a "transparent" computer, the flow of information which circulates between a primary school class and its environment. This means that a computer "tool" intended for word processing and image creation is being used in a school context, without the children needing to know how it works or to be familiar with any programming language. The original feature of this work is that it is relatively long-term (two school years), and includes studies ranging from observing child behaviour in the classroom to a sociological survey of families.

Various soft- and hardware is used in different educational situations. Results of the study will be published in the form of a research report in 1988.

Bibliography

J-F BOUDINOT, J PERRIAULT, L'école point d'accumulation et de redistri-
bution d'une culture locale élaborée par les enfants. Etude du rôle
d'un ordinateur-outil. In: Réseau Communication-Technologie-Société,
Revue du CNET/PA/UST, No. 13, Paris, 1985, pp. 27-38.

1.5.4   INFORMATION TECHNOLOGY

If there is one form of technology that foreigners envy us for, it is the
"Minitel", the most conspicuous - but not the only - aspect of French
information technology; to this, interactive view-data, video communication
("visio-phone", optical fibre) and computer-assisted conferences should also
be added.

The latter is not currently the subject of INRP-research, although we are
following with interest the experiments carried out by some American universities and the   British Open University.

1.5.4.1   *Educational application of video communication*

Research is based on the setting up of an experimental optical fibre and
"visiophone" network in Biarritz (South-West France), installed in peoples'
homes, which provides standard telephone services plus the following:
access to IT-services (alpha-numeric keyboard), transmission and exchange
of video images by means of an integrated mobile camera and/or other
plug-in audiovidual equipment video recorders, video discs, etc.). The
work is being carried out in conjunction with the FOEB-project (optical
fibre education), run by the Bordeaux Regional Centre for Educational
Documentation. It consists of installing a home assistance service, for
secondary school pupils, using a video-phone, and exploring educational
uses of the network in a teaching situation. The teachers/associates took
part in a technical and methodological training programme from 1984-1985.

The research aims to explore educational uses of the video-communication
network, conditions and possibilities of mastering this technology for both
pupils and teachers, and possible modifications of the teachers' educational
role.

A research report will be published in 1988.

1.5.4.2   *Production and testing of educational IT-services using viewdata*

This work was carried out following a request from the Ministry of Education for the INRP to participate in an experiment using interactive viewdata
for educational purposes in 21 educational institutions in the Vélizy-Ver-
sailles region (near Paris), between 1982 and 1984.

The original feature of "Télétel" (which has no doubt led to its international
reputation), is two-fold: first, it is free, and second, it is easy to use.

Any telephone subscriber can have a Minitel installed, free of charge, i.e. a terminal which is easy to use and gives access (rarely free!) to numerous sources of information, at various distances, and which can be constantly brought up to date. The extraordinary boom in services available has far exceeded the expectations of the Telecommunications Department. The French education system cannot ignore this veritable revolution in access to information, nor leave it to private companies to develop products intended for learners (computer-assisted learning, "emergency" services such as "Maths Secours", tuition for exams ...).

The experiment carried out in Vélizy revealed the complexity of preparing educational services on viewdata: specific, very restricted vocabulary, which could impoverish the language, and the need for a minimal knowledge of documentary techniques, not to mention access problems: additional costs for schools, the different attitudes of the authorities and even teachers, lack (or sometimes poor quality) of services proposed. The current research has several aims:

- Teacher training, with a view to manufacturing educational products;

- Creation of software to help teachers compose viewdata screens;

- Development of local data banks (primary: the Palace of Versailles; secondary: the geography, history and economy of a region; technical education: on dipoles).

Bibliography

P GUIHOT, Expérimentation du vidéotex-télétel dans 21 établissements de la région de Vélizy. Rapport de recherche, INRP, 1984.

P GUIHOT, Télématique et éducation, l'expériment Vélizy. Rapport intermédiaire, INRP, 1985.

P GUIHOT, A videotex integrating experiment in French school. In: Videotex Canada, July/August, 1985.

Les usages éducatifs de la télématique. Collection Rencontres Pédagogiques, INRP, 1987.

1.5.5   AUDIOVISUAL

Under this heading, I intend to regroup various research on "images" from television to the video-disc.

However, I would like to mention the development of new audiovisual practices in schools due to portable video equipment being made available to classes. This phenomenon could go unnoticed, overshadowed by currently fashionable technology (microcomputers, IT, video-disc), but it deserves to be mentioned because it enables pupils who easily adapt to using new technology to learn production and creation techniques.

The INRP has begun an enquiry in schools on cinema and audiovisual education; the first results indicate that research is needed to better understand this phenomenon.

1.5.5.1 *Television, a social and educational fact*

Various current researhch projects are concerned with relations between education, on the one hand, and communication and information through television, on the other. Most of the work does not particularly concentrate on interactive use of audiovisual techniques in classes. However, I would like to mention some aspects of the work which could interest the reader who is curious to know more about the relationship between television and children:

(a) <u>Television and children's understanding of events and of things foreign</u>

This work was done with children aged 9-10 from two standpoints, one mainly sociological and the other mainly psychological. A series of four videos has been produced with the title "La classe face à l'évenement" (B. Poirier, 1985). A research report will be published in 1988.

(b) <u>Study on effects of certain types of television output</u>

American and French series, historical television films, popular science programmes, on the social and cultural perception of young people aged 11 to 17.

The final report on this research will be published in 1988.

Bibliography

B CHAPELAIN, Dallas et son succès. In: Etat du Monde, 1982, Paris, February, 1983.

B CHAPELAIN, Télévision. In: Revue Cadre de Vie, January/February, 1984.

B CHAPELAIN, Effets des feuilletons Américains et Français sur les représentations sociales et culturelles chez les jeunes 11-17 ans. Communication of the Colloquy "Enseignement Audiovisuel et Cinéma", Paris, February, 1986.

B CHAPELAIN, Les jeunes et les feuilletons. In: Revue APTE, January, 1987.

B CHAPELAIN, La fiction à la croisée de la vulgarisation scientifique: jeunes, télévision et histoire(s). Rapport de recherche, INRP, 1987.

(c) <u>How children (aged 7-9) learn from television</u>

Using an original audio-visual programme on energy, seen in different everyday situations, this research aimed to observe and reveal possibilities of semiological and/or cognitive activity in children, in relation to various reception contexts in the educational environment.

A research report will be published in 1988.

(d) <u>How memory aids influence the capacity of children (aged 7 to 8) to recall television documentaries and fiction</u>

A research report will be published at the beginning of 1989.

1.5.5.2 *Educational uses of the video-disc*

Carried out at the same time as the Ministry of Education's installation of video-players in some 100 schools, the research's aim is to study the integration of this medium into secondary education; its subsidiary aim is the creation of material and methodology needed for educational purposes.

The research team have produced a video-disc called BASILIC "Base d'images à lecture interactive"), and a series of software, including one "VIDEOBASE" - for management, all of which have been distributed in the schools concerned.

The first part of the research studies pupils using tools that could be applied to several different subjects (French, natural science, history, languages, art).

Pupils' activities (word processing with image inserts, creation of relational bases with fixed images and motion sequences) have proved useful in producing objects that can be re-used by other pupils or teachers

The second stage of the research uses more sophisticated material, not provided by the Ministery of Education (PC-compatible and Macintosh), and is therefore speculative. One of the team's software programmes, using the "castles of the Loire" video-disc, allows an interactive visit to the castles with access to textual and graphic information. Using the same video-disc, the team has developed a written application in PROLOG to process the French royal family tree.

In the field of educational use of video-discs, creating an environment which allows the tools' interaction possibilities to be fully exploited, involves a didactic study which is both innovative and in-depth.

During the experiments, the following were compared: the advantage of using one screen (overlaying) or two, collaboration with researchers and teams from other public or private laboratories, in France or abroad.

1.5.5.3 *Educational use of information technology*

In the first place, two aspects will be considered:

- Development of data banks and simplifying access to them;

- Development of inter-personal dialogue and data exchange networks.

Research will be concerned with:

- Observing the behaviour of children, adolescents and teachers in relation to these two phenomena, depending upon various criteria (age, sex, social background ...);

- Study of assimilation of new technology in an education environment;

- Analysis of changes in the hierarchic structure due to new methods of access to information, self-assessment, peer assessment, ...;

- Creation of educational items and situations allowing integration of this new potential into schools and other establishments;

- Effects of technology on the content of certain syllabuses and/or curricula.

1.5.5.4 *Technological imagery in education*

This third field of research aims to study different representation methods available to the learner, various "subjects" (academic attainment, social life, family structure, technological tools, ...) through new technology as a new means of communication.

The "image" aspect in particular will be taken into account and analysed:

- Relations: representations/types of "image": photographs, films, videos, synthetic image ...,

- Creation/production/assimilation,

- Relations: image/sound/text,

- The role of "images" in various fields of education, artistic or otherwise,

- Recognition of new aptitudes and knowledge linked to the technology.

1.5.6 INTERACTIVE LEARNING OR LEARNING INTERACTION?

Some remarks and questions to reconcile new technology and life in the classroom.

Remark 1

Decades of educational tradition are not made to disappear simply by installing machines to produce images, sounds, texts, exercises, and even information, in the classroom.

Question 1

What training - and in what form - should be given to future teachers in universities and teacher training centres, so that in schools of all kinds, diagram A can be replaced by diagram B?

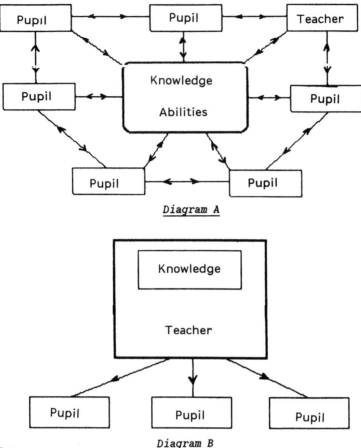

Diagram A

Diagram B

Remark 2

At a professional level, new technology forms part of the equipment needed to develop science and technology: mathematicians solve differential equations in a different way with a computer, meteorologists make forecasts in a different way with the help of satellites, economists make forecasts in a different way with computerised data banks; and they all do new thins with the technology available.

Question 2

On what condition(s) will teachers and those responsible for syllabuses and curricula accept new technology, not as a means of reproducing traditional educational contents and method, but as a means of teaching the same, or different, contents in a different way? How will new skills and performance directly linked to mastering the new technology be taken into account in learner development?

Remark 3

New technology increases human possibilities for accessing, storing, and processing of information.

Are educational systems sufficiently aware of this phenomenon to provide learners with the means to master new techniques? Are educational systems up to teaching interaction, so that learners can benefit from interactive learning, and not become dependent upon a new intermediary, i.e., not the person who has the information (Diagram A), but the one who knows how to gain access to the information (Diagram C)?

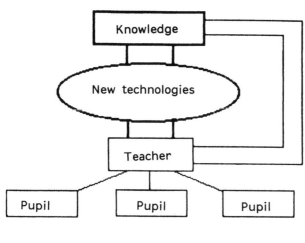

Diagram C

Remark 4

Artificial intelligence has begun to bear fruit in industry, and will soon do so in the tertiary sector.

Question 4

Are educationers prepared to make use of work on artificial intellience to improve learners' access to the new sources of information and training?
Are the ministerial authorities concerned prepared to encourage installation of top-of-the-range technological products in schools, and provide adequate training for the teachers?

Remark 5

Access to a computer limited to a few hours a year per pupil; a "minitel" reserved for teachers, documentation, or indeed the head of the school:

Question 5

This being so, can we really speak of interactive learning through new technology, with a school context? Won't this situation tend to emphasise the differences in performance at school of children from under-privileged and those from privileged backgrounds?

1.6 INTERACTIVE LEARNING AND NEW TECHNOLOGIES IN THE FEDERAL REPUBLIC OF GERMANY

by
Dr. PETER MICHAEL FISCHER and Prof. Dr. HEINZ MANDL,
German Institute for Distant Studies, Tübingen

1.6.1 SUMMARY

The present article, on the one hand, gives an outline of the state of art in educational computer use in the Federal Republic of Germany. Educational interactive computer programmes from simple drill and practice, programmes running on small cheap machines, up to artificial intelligence based 'microworlds' running until now only on mainframes are sketched. This overview also argues that sophisticated audiovisual interactive new technologies can properly be used, even on the small machines available in schools, if the content materials and the flow of control given to the learner are carefully designed and are based on principles of cognitive science.
On the other hand, the spectrum of possible meanings and interpretations given to the concepts 'interaction' and 'interactive' is elaborated and prototypically illustrated by typical programmes of each kind.

1.6.2 INTRODUCTION

All learning is interactive. There is no human or even animal learning without an interaction with a social agent of learning who informs about the results of learning or at least about some form of informative event (implicit feedback), e.g. by observing the results and outcomes of own doing and acting. Interactive learning may thus be roughly equated with learning by dialogue or learning by feedback. Dialogues or feedback delivered from any of the new learning technologies in the course of learning may mimic or simulate human real life dialogues. These may be defined as an instance of a flexible and unrestricted two-way, two-channel communication, switching back and forth between a sender and a receiver who continually change their roles.

A critical review and inventory of current developments in the realm of interactive learning with the new technologies needs a caveat and a guiding principle however: 'Dialogic learning', 'interactive learning' or 'learning by feedback' are not means and ends in themselves, but rather a modal category by which superordinate pedagogic goals can be reached - or be reached somehow better. The adequacy of interactive learning for the paramount pedagogic goal to be reached depends upon several crucial components and their interactions. To enumerate just a few of these: The domain in question, related pedagogic and tutorial rationales, philosophies and strategies, the target learning goals - from elementary, factual knowledge or basic skill training to the acquisition of complex knowledge, from perceiving, acting, knowing and comprehending to theorising and gain of insight - they all determine whether interactive learning in the one or the other sense is adequate or, in a given implementation, is adequately realised.

Norms and standards to evaluate a given teaching/learning dialogue must be linked with the pedagogic and curricular context in which the dialogue takes place. A taxonomic scheme to fix the position, and to critically review current developments in the realm of interactive learning, has to be sensitive to pedagogic and tutorial rationales, philosophies and strategies, as well as to the target learning goals.

1.6.3   A TAXONOMY OF COMPUTER-ASSISTED TEACHING/LEARNING

Basically one can differentiate between what has to be learned with respect to what has to be taught: is the focus directed to the mediation and teaching of a <u>content area</u> in respect to a <u>given domain</u> directed to the basic, elementary factual level, or to the higher order level of comprehension and understanding of complex knowledge? Is one willing to lay more stress on the <u>activity per se</u>, that is, the <u>learner's intelligent activities</u> and the process of his problem solving or thinking? The answers to these questions govern whether one chooses either a <u>drill and practice</u> and/or <u>tutorial approach</u> or a <u>learning-by-doing, heuristic approach</u> by simulations and/or Microworlds. Any further dealing with the applicability and use of the new technologies as means to aid learning is dependent upon this basic dichotomy (c.f. Figure 1).

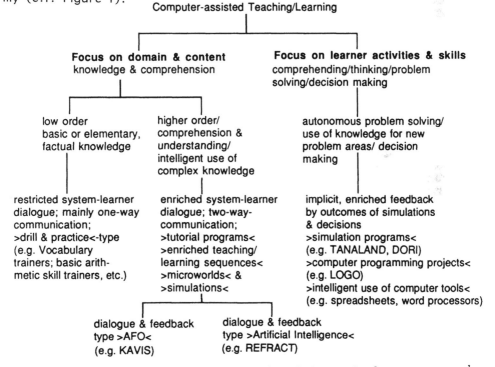

Figure 1: *Basic dichotomy between domain/knowledge-centred programmes and process/activity-centred programmes*

1.6.4 COMPUTER-ASSISTED TEACHING OF DOMAIN-RELATED KNOWLEDGE

1.6.4.1 *Teaching of elementary factual knowledge by drill and practice*

In a sense drill and practice-programmes are a sequel of Skinnerian teaching/learning machines. 'Dialogue' in simple Computer Assisted Instruction programmes (CAI) is restricted to short comments stating that the learner's input was 'true' or 'false'. Learner responses, if true, are followed by a 'reinforcing event' (e.g., a pleasing sound, a jumping lion, and the like).

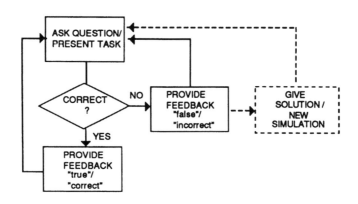

Figure 2: 'Prototype' of drill and practice-programme

There is not much 'intelligence' implemented in the less sophisticated programmes of this type. Foppa (1968) has criticised CAI-programmes of Skinnerian heritage by the following arguments: analogies of conditioning and learning are rather vague. CAI is neither pure classical conditioning nor behaviour shaping. 'Responses' to be learned (and to be reinforced) are not exhausted by pressing a bar, but rather consist in the discrimination and selection of a target answer, which has to be matched with the original question. Not an operant, emitted response is shaped but an elicited one. Besides that there is no dealing with extinction phenomena, which one has to be aware of when treating one-trial learning. Most important is the following criticism: in a typical drill and practice-task the feedback 'False!' following the selection of a false answer does not help much. There are too many possible answers left out, which by 'trial and error' might result in further series of 'False!' statements until - by chance - the correct answer is hit. If we think about both the informing and the reinforcing qualities of feedback, then continued falses will yield in dropping motivation, and possibly in lowered self-esteem of the learner.

Even though there are weaknesses, drill and practice-programmes are advantageous in some regards: they are running on the small, cheap machines widespread in our schools, do not require much memory or numeric power, are written in commonly used computer languages, and are thus modifiable by the individual teacher. They are economic in the sense that they could swiftly be developed without much expenditure to designing and programming. There are numerous drill and practice-programmes in circulation, no less than 124 pages list programmes developed just for the Apple computer. A typical drill and practice-programme may be illustrated by the following example:

```
•••••••••••••••••••••••••••••••••••••••••
* Chemistry Elements/Symbols Test *
•••••••••••••••••••••••••••••••••••••••••
```

You are getting 10 seconds to ENTER the wanted chemistry symbol by the keyboard.

Please ENTER only Uppercase !

Hit RETURN when ready !

Computer Display/*Comment*	Learner Input
What is the symbol for calcium ? *You are thinking too slowly !*	CA
What is the symbol for copper ? *Great !*	CU
What is the symbol for caesium ? *False ! The symbol for caesium is CS !*	C
What is the symbol for krypton ? *False ! The symbol for caesium is KR !*	CR
What is the symbol for phosphor ? *False ! The symbol for phosphor is P !*	PH
What is the symbol for strontium ? *False ! the symbol for strontium is SR !*	ST
What is the symbol for tantal ? *Yeah !*	TA
What is the symbol for actinium ? *True !*	AC
What is the symbol for carbon ? *Yeah !*	C
What is the symbol for germanium ? *Great !*	GE

EVALUATION

True	five times
False	four times
Too slow	1 time
You have needed	51.9 seconds
Your score is	134

Figure 3: Part of a drill and practice chemistry-programme by F. Kappenberg©, Chemistry Programme Library, Chemistry Lab. Dr. Flad, Stuttgart, Germany

More elaborated drill and practice-programmes provide a more thoroughly designed, expanded feedback. Different feedback may be given according to the number of erroneous trials. Again this may be illustrated by an example:

```
••••••••••••••••••••••••••••••••••••••••••••••
*  Klett© Training-Software ALGEBRA I  *
••••••••••••••••••••••••••••••••••••••••••••••
```

Computer Display (*Comments & feedbacks in italics*)

RULE:
Extending means multiplying the numerator and the denominator by the same factor for extension.

In order to extend 2/3 to ?/15 one has to multiply the numerator by five as the denominator was multiplied by 5.

$$\frac{2}{3} = \frac{2 \times 5}{15} = \frac{10}{15}$$

Task:
$\frac{5}{12}$ extended by 22 = $\frac{?}{?}$ (110
                                          264)

If the response is permanently incorrect the following feedback sequence is displayed:

1st feedback: *Wrong !*

2nd feedback (gives specification): *The nominator and the denominator are to be multiplied by 22 !*

3rd feedback (gives complete explanation of the rule in question): *Extending means multiplying the numerator and the denominator by the same factor for extension.*

In order to extend 2/3 to ?/15 one has to multiply the numerator by five as the denominator was multiplied by 5.

$$\frac{2}{3} = \frac{2 \times 5}{15} = \frac{10}{15}$$

4th feedback : the nominator searched for is displayed inversly or coloured

5th feedback: *False !* by sound and flashing display the attention is drawn to the already filled in nominator

6th feedback: Accompanied by buzzer or sound the denominator is being filled in as well.

Next Task
the difficulty level of the next task is dropped in correspondence with the following learner answers until the rule has been applied correctly, e.g.:

$\frac{1}{2}$ extended by 3 = $\frac{?}{?}$ resp. $\frac{1}{1}$ extended by 3 = $\frac{?}{?}$

Figure 4: A more elaborated drill and practice-programme

Learners guided by such a programme architecture might suffer from less boredom and monotony, and might profit more than from the most simple programmes of this type. Nevertheless, 'dialogue' and communication between the system and the learner is mainly one-way, one-channel; the learner is restricted on narrow circumscribed responses by the keyboard. The feedback transports, as is the case with the example given above, just the same information as is contained in the teaching/instruction part, where the rule is formulated.

1.6.4.2 *Teaching of complex knowledge by more elaborated AFO- or ICAI-type programmes*

1.6.4.2.1 The differentiation between AFO and ICA

Feedback principles are central components in current Intelligent Computer Assisted Instruction (ICAI). In order to name a tutor system 'intelligent', its quality and ability to perform on-line diagnosing and modelling for the learner is crucial.

According to Wexler (1970), the former CAI systems were Ad-hoc Frame-Oriented (AFO) or "generative", because the ability to react of the learning device was restricted to a fixed event-space anticipated at the time of system construction.

Figure 5 illustrates a 'prototypical' AFO architecture as contrasted with an ICAI architecture (see Figure 6 below).

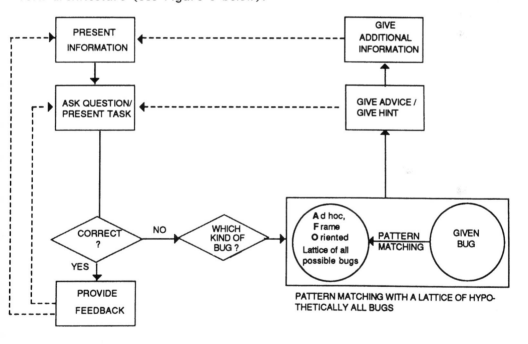

Figure 5: 'Prototype' of an AFO-teaching/learning system

59

AFO-teaching/learning systems may be designed most sophistically, but suffer from their 'fixed' error-detection/'bug diagnosis' which assigns error-sensitive, differential feedbacks. At the other side, their fixed event-space anticipated at the time of the system construction is a must when audio-visual feedback information is provided. The 'bottleneck' of time consuming designing of feedback-screenplays and their video-production restricts audiovisual teaching/learning systems to the AFO type mentioned.

In contrast recent intelligent tutoring systems or ICAI systems are claimed to be flexible with respect to a wide range of unforeseen events. Their flexibility, which may be equated to the definition of their "intelligence", is based on their potency to diagnose. Koffman and Blount (1975) define the intelligence of a system according to how well the system is able to reconstruct and hypothesise the sources of difficulty for a learner from the learner's recent learning history with the system (c.f. Figure 6).

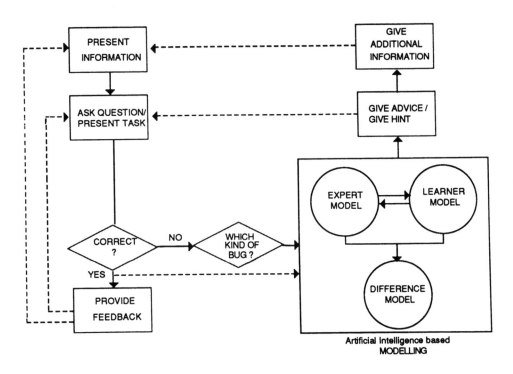

Figure 6: 'Prototype' of an ICAI architecture

One must be able to precisely map and analyse the problem the learner is confronted with, to make a tutorial intervention or instruction "intelligent" in the very sense.. This implies either an 'overlay' from which the bug is diagnosed or a normative model of idealised, bug-free understanding of the domain in question or a topic derived from that domain ("expert model").

As a further implication, such a normative or "expert model" needs a deeper insight into the concepts and relations within the domain (conceptual network), and an insight into the nature of the cognitive processes necessary to comprehend the domain. To estimate the current difference between expert knowledge and the learner's actual state of comprehension, a valid model of the learner's present state of knowledge and/or comprehension has to be constructed. Series of content-valid diagnostic tasks and questions are needed. To be able to differentiate (and eliminate) unsystematic "noise" from systematic effects in the learner data some reliability is needed. Only by a series of diagnostic tasks involving some retesting can a valid, noise-free learner model be obtained. Only then, by comparing learner and expert model, can a valid and reliable difference model evolve which then allows a problem-sensitive tutorial intervention.

1.6.4.2.2   KAVIS© as an example of an AFO architecture

The teaching/learning system KAVIS© (= Knowledge Acquisition Video Instruction System) aims at the development and scientific evaluation of a computer-assisted feedback-based audiovisual instruction system which is founded on the ground of cognitive science and cognitivistic learning theories. Knowledge acquisition, from basis, elementary facts to the understanding of cohesive, complex knowledge should be aided by concomitant on-line feedback. Corresponding with different phases in the process of knowledge acquisition and comprehension from information intake to the acquisition of complex knowledge and understanding, we have to be aware of concomitant processes also on the affective level. To be sensitive also for affective co-processes, both our feedback was based on the following rationale: learning in the beginning when the new material or information to be learned is still unfamiliar, lays a heavy burden on the learner, if not even 'stress'. If he then is confronted with a series of negative feedbacks after failure (which is more likely than 'hits'), his motivation and/or self-esteem will drop. He then is affectively conditioned to interpret even informative feedback, as an aversive or punishing event. Only after some mastery has taken place, and after the learner is more familiar with the material, should he be confronted with corrective feedback (Stapf, Fischer & Degner, 1986). The same conditioning phase, again, will appear when he then progresses from elementary facts to higher order, complex understanding, especially if comprehension required is based on valid inferences. Again, a series of negative feedbacks after failure may result in a drop of motivation and/or lowered self-esteem. If feedback would be restricted to a solely informing base, this would not offer much help. To be really aiding, it must also give substantial help, be it in the form of an explanation, or be it in the form of an intervention which provides the learner with a means to monitor and diagnose his failure and to recover from error. We therefore provided two kinds of feedback:

-   A meta operational, controlling and guiding feedback ('PROLOGue') informing him about the logical status of his response - there were major types of error category defined as either 'omissions', 'intrusions', or 'omissions and intrusions' - which provided error-type information to help the learner autonomously correct his response combined with a linkage to the topic in question;

-   An exclusively informing audiovisual feedback ('AV'), presenting advance organisers to the topic in question, and explaining the topic by compressed audiovisual and textual information. A further source of information required for control of learning is the DIAGnostic protocol analysing learner bugs and giving advice about how to proceed (e.g. repeat a topic in case of fatal error, re-answer questions in case of less serious bugs).

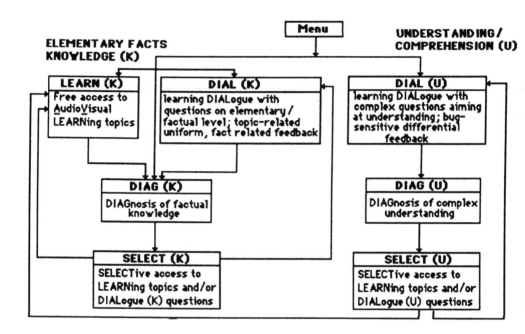

Figure 7: Architecture of the audiovisual, feedback-based teaching/learning system KAVIS©

Figure 7 shows the learner is first given an opportunity to choose the mode of his learning. He can either start with information intake (LEARN) or with an entrance DIAGnosis to get informed by a printed protocol about his present state of knowledge with respect to weak points where he should compensate for knowledge deficits. If he chooses LEARN, he can select from a table of topics with which of three main topics he will start. He then can proceed in a linear order topic by topic or can switch to and fro between thematically grouped blocks of main topics. If he leaves any of the thematically grouped blocks of sub-topics, he is given an opportunity to answer DIALogue questions related to the previous block (DIAL1). Since he has little familiarity with the new material in this early phase of learning, and since he could be negatively reinforced by 'misses', only instrumental, informative AV-feedback is given here. After a substantial progress he then can proceed to a DIAGnosis of his actual learning state, which is followed by a diagnostic protocol. This protocol marks both topics where he should better go thoroughly through again (i.e. LEARN again), and topics where a new DIALogue (i.e. questioning) is sufficient. He then can correct his shortcomings and compensate for his previous weak points by SELECTively choosing either LEARNing topics or DIALogue questions. If he is then able to pass a threshold - mandatory elementary knowledge required for comprehension - he then can proceed to the comprehension and understanding part of the system, where he is confronted with a series of complex questions both in sequential (DIAL2seq.) and randomised order (DIAL2rand.). As mentioned above, differential feedback corresponding the kind and logical status of the learner's bugs is delivered here (either as PROLOGue only, AV only, or PROLOGue & AV combined feedback).

The system makes use of a computer-controlled VHS-video-player with intelligent remote control, and delivers PROLOGue and AV feedbacks immediately after learner inputs. Feedbacks containing an audiovisual component, follow very precisely with a maximal delay of 20 seconds.

To evaluate the efficiency of bug-sensitive, differential on-line feedback, three combinations of feedbacks were tested with 96 uppersecondary students selected according to their high and low prior knowledge (as indicated by a pretest); PROLOGue only vs. AV only vs. PROLOGue & AV.

To control for feedback interpretation effects with respect to the intended affective neutrality of error-dependent feedback, a questionnaire related to feedback use and interpretation was administered before treatment, during treatment (at the point between knowledge and comprehension, that is after DIAG), and after ending treatment.

Learners with poor prior knowledge could be aided optimally by the AV only feedback (being less able to interpret and use the more demanding PROLOGues and being possibly less able to control their learning meta-operationally). Learners with high prior knowledge rather profited from the meta-operational, controlling and guiding feedback information as is contained in the PROLOGue only, PROLOGue and AV feedback information.

With regard to affective co-processes of knowledge acquisition and understanding, the subjective learning experiences of the students and their interpretation of feedback as informative vs. reinforcing/aversive/punishing events could be influenced positively: Instead of interpreting and experiencing learning control as a source of punishment and possibly subsequent avoidance, students recognised and used feedback as a necessary tool to learn from bugs.

Since the system allows for a fine-grain on-line recording of any learner activity at any given point of time (3660 data were recorded per learner containing even latencies for any given alternative answer), the system is open for a scientifically and empirically based modification of tasks, feedback-dialogues, and video-tapes. Only after a thorough re-analysis of the curriculum, a laser-videodisc should be produced - at least until laser-videodiscs are not erasable. In a sense, systems of a certain degree of sophistication can lay the empirical grounds for the modelling-component in ICAI-systems: on-line recording of erroneous learner inputs can directly be used when designing a "bug-lattice", leading to the construction of valid and reliable diagnostic test items.

At least, until now, 'authoring systems', which can easily be used, have not been available. Since most programming languages, at least if the programmer uses them modularly and systematically, are more powerful and not harder to learn than 'authoring languages or environments', we encourage teachers to use available and user modifiable tools, rather than to cope with the 'current state of art' of authoring systems.

1.6.4.2.3   Other types of AFO and non-ICAI architecture

Some simulation programmes offer the opportunity to input numeric values calculated or estimated from physics or chemistry laws. After user input, the computer simulates and outputs the result in text, graphics or even animated graphics. (An example are the 'Programmes for Physics Teaching' by H. Haertel, Institute for Science Education, IPN, Kiel.) Feedback in a sense here is substituted by observing the results and outcomes of one's experimentation or hypothesising; learning thus is guided analogous to Thorndike's law of effect in its truest sense.

A recent development of a sophisticated and complex simulation package is SCHULIS© by the Gesellschaft für Mathematik und Datenverarbeitung (GMD). SCHULIS© is a computer-assisted simulation tool which can be applied to numerous models in the natural sciences. Testing of models dealing with model assumptions is done uniformly and consistently by one user-interface, even if the content of the models can vary between different field of natural science. SCHULIS© consists of a simulation tool apart from concrete models, and database/library of models, model data, and model equations. Currently, models from biology, ecology, and physiology are released.

1.6.4.2.4   REFRACT as an example of an ICAI architecture

REFRACT by P. Reimann, University of Freiburg, is based on the 'microworld' approach (c.f. Lawler, 1982). A microworld is an interactive learning environment which gives the learner a facility to learn by the testing of hypotheses by conjectures and refutations (learning by discovery in a Popperian sense). Starting point is a confrontation with a problem from a narrowly circumscribed domain which is intended to induce a conflict between prior knowledge, common-sense-models or naive, pre-scientific assumptions on the part of the learner. If he experiences a gap between his earlier assumptions and the obvious 'anomaly' imposed by the problem, he is motivated to settle the conflict. He is intended to test hypotheses step by step, then, in a cyclic problem-solving approach.

REFRACT mediates knowledge from the domain of geometrical optics. Its knowledge base is the refraction of light beams at the surface of two different media, and the reflection characteristics of thin lenses. By experimenting and manipulating with the microworld REFRACT, the learner is expected to get a better understanding of the laws of refraction, and to acquire metacognitive strategies by conjectures and refutations.

The instructional sequence implemented in REFRACT is the following:

(1)   The learner has to choose among media, the shape of the optical surface between air and the medium chosen (convex/concave), the distance between the input ray and the medium on the optical axis, and the angle between optical axis and input beam.

(2) The learner then formulates a hypothesis about the refraction of the input beam with different degrees of precision. He can either spefciy an area on the computer display he supposes to be the location of the refracted beam, as a kind of a somewhat general prediction. Or he can draw the refracted beam on the display, as a kind of a more precise prediction. Or he is allowed to formulate his prediction in quantitative, numeric terms specifying the angle of refraction (see Figure 8).

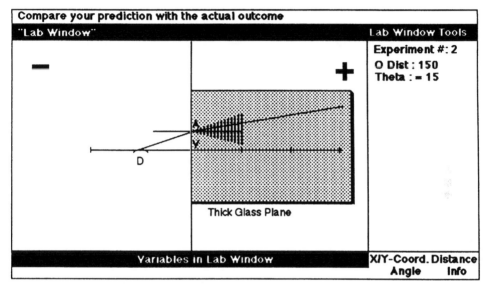

REFRACT-Experiment with a Thick Glass Plane

Figure 8: 'Manipulation' and 'simulation' window in REFRACT

(3) The programme implemented in REFRACT gives information in the form of a graphic representation of the correct results.

(4) The learner processes this information by relating his own prediction to the true, observed results. A 'notebook'-window assists this processing by a panel which records the learner's predictions and the true outcomes of the experiments. The learner, thus, is given an opportunity to trace his progress and to re-analyse and reconstruct his previous experiments.

REFRACT thus is a prototype of the currently most elaborated ICAI systems. The spread and distribution of such elaborated programmes has been

65

restricted to the artificial intelligence/cognitive science lab until now, since they can be run only on powerful but expensive LISP-machines, with at least 1.5 Mb RAM and 53 Mb hard-disc capacity with high resolution graphic screens like the Xerox X-1108.

There is good hope, however, that the powerful new microcomputers forth-coming in the next months (e.g. Apple's Macintosh II) will break this barrier.

1.6.5 *Computer-assisted 'hands on' learning of intellectual skills*

There are to main types of computer-assisted 'hands on' learning paradigms of intellectual skills: simulation programmes/environments aiming at the analytic, and hypothesis testing skills of learning and complex decision making. Simulation programmes from biology/ecology (e.g. H. Bossell's DYSYS© simulation kit for a series of problems), and problem-solving pro-grammes (e.g. D. Doerner's MORO, TANALAND, DORI simulation), which require complex testing and decision-making are representatives of this type. The other 'hands on' approach to foster intellectual skills is computer programming itself, e.g. in LOGO. Investigations concerned with LOGO programming were mainly focussed at the interactive programming approach. Empirical evaluations did not show substantial far-transfer effects on pro-blem solving or algorithmic thinking and analytic skills, at least beyond the narrow circumscribed area of programming skill (near-transfer).

To complete the short description of prototypes for the different approaches to dialogic, computer-based learning, a short sketch of a derivative of D. Doerner's DORI-programme is given.

Simulation programmes of the problem-solving type confront the learner with a scenario in narrative form, containing all information required for doing the simulation. Usually the learner is induced to play the role of a develop-ment aid volunteer in an under-developed country. He has a certain amount of money available to pay the current expenses of his activities. The lear-ner then can choose between a couple of steps, actions, and measures to intervent (c.f. Figure 9).

Dependent upon the sophistication of the programme developer, and the power of the computer used, the network of interactions between the para-meters of the simulation can be very complex (see Figure 10).

Slimmed versions of simulation programmes are running even on smaller machines and are widely used in education.

With respect to transfer effects of 'microworld' manipulation/simulation on general thinking, there is only scarce, most intuitive proof as is the case with LOGO programming. This is partially due to a lack of empirical long-term, time-series studies, especially with respect to long-lasting effects of such training. This is also due to a lack of clear and conclusive concepts and derived testable near-transfer and far-transfer indicators of 'positive transfer effects on general thinking', which are mostly stated in an enthu-siastic phraseology, and are claimed to exist but which are not formulated in a methodologically adequate way. There are two sides to everything: claiming to foster thinking and problem-solving skills by educational use of

66

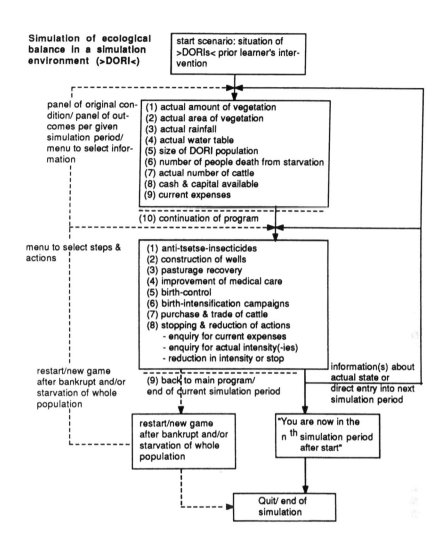

Simulation of ecological balance in a simulation environment (>DORI<)

start scenario: situation of >DORIs< prior learner's intervention

panel of original condition/ panel of outcomes per given simulation period/ menu to select information

(1) actual amount of vegetation
(2) actual area of vegetation
(3) actual rainfall
(4) actual water table
(5) size of DORI population
(6) number of people death from starvation
(7) actual number of cattle
(8) cash & capital available
(9) current expenses

(10) continuation of program

menu to select steps & actions

(1) anti-tsetse-insecticides
(2) construction of wells
(3) pasturage recovery
(4) improvement of medical care
(5) birth-control
(6) birth-intensification campaigns
(7) purchase & trade of cattle
(8) stopping & reduction of actions
   - enquiry for current expenses
   - enquiry for actual intensity(-ies)
   - reduction in intensity or stop

restart/new game after bankrupt and/or starvation of whole population

(9) back to main program/ end of current simulation period

information(s) about actual state or direct entry into next simulation period

restart/new game after bankrupt and/or starvation of whole population

"You are now in the nth simulation period after start"

Quit/ end of simulation

Figure 9: Flow-chart of a problem-solving simulation programme

simulation programmes, and instructing children in programming is just one side of the coin; proving that this is, in fact, the case, means that one has to delineate where one expects such effects and how they can be reliably measured. Cogent indicators for such effects have scarcely been defined, at least until now. Only if one could show that there is, say, a specific near-transfer long-term effect, e.g. on the programming skills of today's LOGO-child in the nineties, when it has become a software specialist, would some of today's great hopes in LOGO be justified.

Network of interactions between simulation relevant forces & parameters of >DORI<

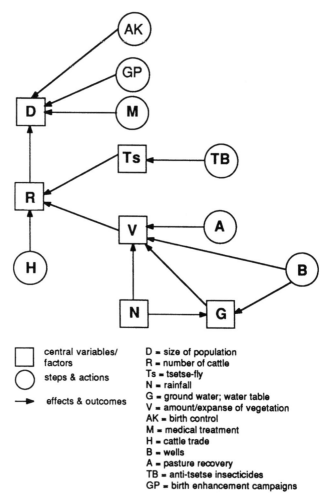

Figure 10: Network of complex interactions in the problem-solving simulation programme DORI

Nevertheless taken all together, the educational use of computers obviously seems promising. If it will flourish even at a fraction of today's hardware efficiency, there is no reason to resign.

1.6.6 BIBLIOGRAPHY

BOSSEL, H. DYSYS^{TM}. Umweltdynamik. 30 Programme für kybernetische Umwelterfahrungen auf jedem BASIC-Rechner. München: te-wi, 1985.

BROWN, J.S., BURTON, R.R. & LARKIN, K.M. Representing and using procedural bugs for educational purposes. Proceedings of the 1977 Annual Conference of the ACM, Seattle, 1977, pp.247-255.

BURTON, R.R. & BROWN, J.S. An investigation of computer coaching for informal learning activities. In: D. SLEEMAN & J.S. BROWN (Eds.), Intelligent Tutoring Systems. New York, NY: Academic Press, 1978, pp.79-98.

FISCHER, P.M. Wissenserwerb mit interaktiven Feedbacksystemen. In: H. MANDL & P.M. FISCHER (Eds.), Lernen mit Dialog mit dem Computer. München, Wien, Baltimore: Urban & Schwarzenberg, 1985, pp.68-82.

FISCHER, P.M., FREY, H.D., & JEUCK, J.J. Entwicklung und Erprobung eines computerunterstützten Video-Instruktionssystems für den naturwissenschaftlichen Unterricht. Forschungsbericht No. 22, Tübingen, Germany: DIFF, June, 1983.

FISCHER, P.M., MANDL, H., FREY, H.D. & JEUCK, J.J. DFG-Projekt: Beeinflussung und Förderung des Wissenserwerbs mit naturwissenschaftlichen AV-Medien bei kontingenter Rückmeldung. Darstellung des Forschungsvorhabens. Forschungsbericht No. 26, Tübingen, Germany: DIFF, May, 1984.

KAPPENBERG, F. Chemie Programmbibliothek. Stuttgart, Chemisches Institut Dr. Flad.

KLETT^{TM} Trainingssoftware Bruchrechnen. Stuttgart: Klett.

KOFFMANN, E.B. & BLOUNT, S.E. Artificial intelligence and automatic programming in CAI. Artificial Intelligence, 1975, 6, pp. 215-234.

LAWLER, R.W. Designing computer microworlds. Byte 1982, 7, pp.138-160.

REIMANN, P. REFRACT. A microworlds program for optical refraction. University of Pittsburgh, Learning Research and Development Center, February, 1986.

STAPF, K.H., FISCHER, P.M. & DEGNER, U. Experimentelle Untersuchungen verschiedener Rückmeldungsmodalitäten beim Lernen. In: K. DAUMENLANG & J. SAUER (Eds.), Aspekte psychologischer Forschung. Festschrift zum 60. Geburtstag von Erwin Roth. Göttingen: Verlag für Psychologie Dr. C.J. Hogrefe, 1986, pp. 221-236.

WEXLER, J.D. Information networks in generative computer-assisted instruction. IEEE Transactions on Man-Machine Systems 1970, 11, pp. 181-190.

1.7 INTERACTIVE LEARNING WITH NEW TECHNOLOGIES; WHEN WILL IT BE SUCCESSFUL?

by
PLØN W. VERHAGEN,
University of Twente, The Netherlands

1.7.1   SUMMARY

New media and information technologies (NMIT) are gradually penetrating modern society. As a cultural phenomenon, they affect life in schools as well. The question arises: Do they have to be adopted into the didactic reality of everyday life? In this contribution it is contended that the answer will be 'yes'. There are, however, some conditions to be fulfilled:

Knowledge and skills with respect to NMIT should be trained. Teacher training should be aimed at the integration of NMIT in the curriculum. Instructional designers, computer specialists, audiovidual designers and other people involved inside and outside the schools should be trained with respect to their desired contribution. Clear roles in this respect have to be defined for each professional. 'Co-operation' is a second keyword next to 'training'.

Research, development, and the implementation of applications have to be distinguished in time and space. Research in the laboratory has to yield results for future developments. Development of applications has to take place in a co-operation between institutions outside the schools and teachers in the schools. The large scale implementation of tested applications is a matter for the schools only. To work along these lines, a suitable infra-structure has to be set up. Apart from this, infrastructures at school level have to be developed to be capable of managing the available NMIT-re-sources, and to support teacher activities with respect to these media.

The use of NMIT applications is primarily a matter of curriculum develop-ment. On any level, NMIT ought to be approached acoordingly. This is a principle that should guide the decisions with respect to training and infra-structure. The consequences of these conditions for policy making on the macro level are discussed.

Two Dutch initiatives with respect to the introduction of information techno-logy and of new media are briefly described.

1.7.2   PROPERTIES

The "newest" technologies include, at present, interactive video, interactive compact disc, CD-ROM, two-way cable and satellite communication, viewdata and other remote data bank facilities. They combine a selection of the next technical properties:

- (Remote) access to information stored on magnetic tape or disc, or on optical disc;

- (Two-way) communication at a distance;

- Video screen output, mechanical reproduction of sound, or printer output;

- Input by switches, dials, keyboards or devices like touch screens or 'mouse'.

The new technologies form an extension of the range of media already available in education. Cerych and Jallade (1986) take them together with longer existing audiovidual and computer media, referring to them as 'new media and information technologies' (NMIT). According to them, NMIT may be used as:

"a. tools for thinking with
b. tools for organising information
c. means of accessing information
d. means of communicating and processing the 'printed' word
e. means of stimulating and reinforcing learning."
(Cerych and Jallade, 1986, p. 29)

For these purposes, application of NMIT is meaningful if typical NMIT properties are being utilised. Central issues are the use of NMIT to bridge time and distance; to store presentations for repeated use; to make large quantitites of information available with virtually no delay; to manipulate data in a way which is difficult or impossible to carry out by hand. The latter includes the management of the user-system interaction.

The range of NMIT includes the possibility to display high quality video images and computer graphics, either separately or combined onto one screen. Also sound may be added. In principle, all the communicative power of still pictures, sound and moving images may be used by NMIT.

Similarly, the available computing power for NMIT is in principle only limited by the state of the art of computer technology. Applications may be close to common classroom practice or more innovative in nature. The use of the computer as tutor is an example of more traditional use. In this case, the computer is placed in one of the roles of the teacher, for instance, to present exercises, to guide a student through a lesson, or to diagnose some aspects of learning behaviour. The computer may also be used as a tool, for instance, as a calculator, as word processor, or as data bank. The tool concept expands the didactic possibilities of computer use. It gives way for new types of learning activities. Especially the use of data banks discloses large amounts of information which may be accessed by students in a variety of ways. NMIT expand the possibilities further, by adding (audio-) visuals and (distant) communication options.
A recent Dutch survey of possible patterns of use in this respect is to be found in Van der Klauw, Van Meeuwen and Timmermans (1987). Expert systems are becoming of increasing importance, offering the student responsive environments in relation to both the tutor concept and the tool concept.

Taylor (1980), who introduced the idea of the computer as tutor or as tool, also mentioned the computer as tutee. This means the computer as programmable, for instance. The use of computer languages like LOGO or Pascal, are examples of computer use of this type. Learning to program a computer is, of course, one way to use the computer as a "tool for thinking".

The question is: What kind of NMIT facilities do we want to develop? Where and by whom will they be applied? Who are the designers? And last but not least: are the costs affordable?

1.7.3   DESIGNING FOR NMIT

The design of NMIT applications requires contributions from different disciplines to decide what is to be designed, and how. The actual design may be partitioned and take place on different levels outside and inside the classroom. The multidisciplinary nature of NMIT design will be discussed in this section.

Figure 1 lists seven interrelated design variables.

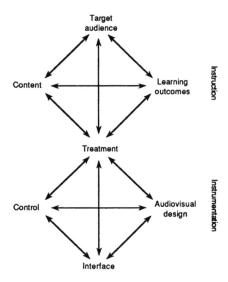

Figure 1: Design variables of NMIT
(adapted from Pals & Verhagen, 1987)

The upper half of the figure represents the instructional variables with respect to organising subject matter into a treatment, to enable certain learners to reach a set of desired learning outcomes. The lower part of the figure represents the instrumentation variables by which the delivery of the instruction takes shape, and the interaction with the learner is managed.

The double arrows symbolise the mutual relationship between the different pairs of variables. Figure 1 is in principle applicable to all instruction. Here it will be elaborated to deal with NMIT in particular.

The choice of media and methods is part of the design decisions to specify the treatment. In that way, NMIT are placed in the centre. The development of NMIT applications, as far as design decisions are to be taken, may be regarded as a two-step procedure. Firstly, the instruction concept is determined, considering target audience, content and learning outcomes in relation to NMIT treatment options, with the instrumentation variables as parameters or constraints. Secondly, the instrumentation in terms of hardware and software is decided upon, with the instruction concept as starting point. These decisions may concern ready-to-use hardware or software. In most cases, however, at least some adaptation will appear to be necessary, and on many occasions, new developments will form a substantial part of the implementation of NMIT in a particular practical situation. In any case, "control" does not only mean the development of the computer software that makes a NMIT application operational in the technical sense, but also the control of the learner-system interaction from an instructional point of view. Providing user-control options, putting questions to learners, giving suitable kinds of feedback and guidance, and registration of user actions for evaluation purposes belong to it. "Audiovisual design" is concerned with decisions about the presentation of images: Video or computer graphics, live pictures or animation, stills or moving video, screen lay-out, etc.. "Interface" is about the physical interaction between the user and the technical system. Ergonomic criteria about, for instance, picture quality apply, but also choices with respect to the input devices like keyboard, mouse, touchscreen, light pen, etc..

The choice or design of NMIT application may take place on different levels. The teacher in the classroom may incorporate available NMIT facilities into his lessons. Political decisions about a national school system may cause a curricular movement which results in an attempt to realise a large scale innovation on the basis of NMIT. In any case, the kind of NMIT use may vary along the range of possibilities which are available as a consequence of the technical and didactical properties mentioned earler. Some characteristic types of use, from which examples exist or which may be developed, are as follows:

-    Database. Computer memories, optical discs, viewdata systems or other widespread networks, may all contain or deliver information which may be searched and used as part of a variety of teaching/learning situations, either in the classroom, guided by the teacher, or individually, at school or at home. The use of database information may be the consequence of specific student assignments, or may take place as a form of open learning.

-    Encyclopedia: for interactive use, or as a structured database for local use. The optical media seem to be promising for this application.

-    Information exchange. Communication between schools and school-related organisations to transfer data and messages between teachers, and for administration purposes. Computer links of any kind are the appropriate media here.

- Distance education. New media may provide a part of the teaching facilities. Telephone consultancy is one example, the use of the home computer as a terminal connected to a host within the teaching organisation another.

- Automated manual. Assembly or disassembly of complicated equipment may be supported and guided by an automatic system, and so is maintenance. Examples are video discs which teach how to build a jet engine, and act as trouble-shooting guide and general manual at the same time (Bayard-White, 1985).

- Computer Aided Learning (CAL) / Computer Based Training (CBT). The applicability of CAL and CBT are well-known. NMIT offer possibilities to expand their range.

A recent publication, edited by Rushby (1987), gives an overview of present thinking about technology in education, offering many ideas and considerations with respect to the use of computers and the newer media.

The question 'which applications have to be developed?' has no direct answer. The possibilities are, technically speaking, so manifold, that the imagination of the user seems to be the limit to a large extent. The applications have to fit in the school culture, however. Here, NMIT skills are involved, and are pedagogical starting points. In this section some remarks with respect to this subject will be made.

Most applications of NMIT will involve the effort of several and sometimes many people. The teacher is just one of them. People outside the classroom take care of the technical infrastructure, the design and production of software, distribution of information and other.

Surveying this, it becomes apparent that the above described design variables refer to people with rather different backgrounds, each of whom contributes to the use of NMIT in education. This may be in a complex project where many people work together, but it may also be in different places and at different points in time, to give a teacher the opportunity to use some ready-to-use software package.

Designing is to be regarded as a craft. In education, theories of instruction, theories about man-computer-interaction, theories about communication or mass communication, may all contribute to the systematic treatment of design problems. Recipes do not exist, however. The skillful designer uses theories as tools for thinking, together with his or her personal experience, full knowledge of the technical possibilities of media which may be used, knowledge of examples, and divergent techniques to generate in difficult cases just that brain wave that puts all components of a certain problem in the right place.

The multidisciplinary nature of NMIT causes some problems. To illustrate this, the design and development with respect to "control" and "audiovisual design" will briefly be described, with interactive video as an example. Two problems occur:

1. People from substantially different professions meet to accomplish one product. They have to learn how to co-operate.

2.   The appreciation of the new technological developments differs from one discipline to another.

Designers of CAL or CBT perceive the video disc just as an extra database of high quality visual images and moving fragments. They tend to continue developing courseware in the way they are used to, incorporating instances from this new database whenever this seems to be feasible. Obviously, this approach widens the scope of CAL/CBT. On every occasion where computer graphics possibilities fall short, high fidelity images may now be made available.

Audiovidual designers, on the other hand, tend to perceive interactive video as a series of video segments under computer control. They continue to develop pieces of linear video, albeit as chapters of a total programme to which stopping points with questions are added, together with a menu to select the traditionally produced video segments in the order the user needs. For some instructional purposes, this approach appears to be a large improvement already, compared to usual video programmes which lack the possibility to go through the subject matter in steps, alternated with exercises and feedback to the learner.

To use the full potential of interactive video, both approaches have somehow to be integrated. Designers and users (teachers and learners) will have to work together on this. With the start of the art in their respective professional fields as part of the background of the designers, the integration will be a gradual process. Design skills are personal skills of the designer. His or her background will determine possible moves into a new area.

This argument applies generally to innovations, and thus for other NMIT as well. Innovations only succeed when they grow roots in the existing culture. Theoretical reasoning, which leads to expectations of a big leap forward because of the wealth of technological opportunities, will in practice always result in little more than small steps (if any). This has frequently been shown in the past. See for instance Unwin (1985), who collected from the young history of educational technology a set of optimistic expectations with regard to programmed instruction, educational television, the use of microcomputers in schools, expert systems, and others. In all cases, the expectations were too high. In section 1.7.4, on implementation, some reasons for this will be given, which have to do with the way in which new ideas were being introduced. At this point, it is sufficient to mention the necessity of linking up with existing practice within the school culture, the CAL/CBT-culture, the audiovisual culture, and - for the upper half of Figure 1 - the worlds of instructional designers and curriculum developers. What this means for policy making, will be discussed in section 1.7.5.

The partitioning of the design, which was mentioned earlier, will take place in two ways: (a) Project teams will divide labour among the specialists from the different disciplines, for instance, to develop an interactive video package; (b) some designs will take place as a series of activities which are not directly connected, for instance, the design of a database system for classroom use as one activity, and the design of the actual use of this system as another one. The answer to the question 'who are the designers?' may therefore be 'anyone, dependent upon the nature of the design problem'. But it has to be clear that designers should stick to their trade.

The computer expert, the teacher, and the audiovisual designer, should not try to do each other's jobs.

In conclusion, NMIT offer a vast array of possible applications in education. Whether these will be elaborated, depends on the question of whether they can be linked up with the present practice. Therefore, teachers (and with them administrators and other staff in the schools) have to recognise the possibilities and incorporate them into their 'personal culture', in concert with the development of pedagogical changes. The upper limit of the speed of the introduction of NMIT, as well as of the level of sophistication of the adopted applications, primarily depends upon the familiarity of teachers with the new attainments, in such a way that they have control over the facilities and are able to design and manage adaptations for their own use. As soon as teachers have become critical and skillful users of software, the demand for quality software will become stronger, while at the same time the term 'quality' will become more precisely defined. Technically speaking, software and hardware are no problem at all, except of course with respect to costs (which will be dealt with in the next section). Technical improvements and new inventions just mean that there is more to choose from.

1.7.4   IMPLEMENTATION

The implementation of new technologies always seems to follow a similar pattern. People from the audiovisual field have been watching the introduction of microcomputers in the schools with some sense of irony. They remember what went wrong with educational television in the sixties. Exactly the same mistakes are being repeated. We may learn some lessons from those days. For this reason, five issues will briefly be addressed.

1.7.4.1   *Buying hardware is always confused with educational improvement*

In the beginning of educational television - at least in the Dutch universities - there seemed to be unlimited funding to appoint studios and to buy equipment. It took years before one became aware of the fact that the actual use of television or video in education was so limited that this bore no relation to the costs. It almost killed the audiovisual centres. The revival of video in the eighties, together with the improvement of the proficiency of media staff and client-centred management, saved these institutions. This, however, is not the point here. The point is that in the beginning the software problem was denied. Buying hardware and hiring staff do not automatically yield desirable educational software that is well implemented. Although this is well-known, the history is repeated with the microcomputers. Tucker (1987) makes this clear on the basis of a survey recently completed for UNESCO. He reports that the great majority of nineteen countries show the same trends. The common factor in the official plans is "the over-riding emphasis on the hardware, to the possible detriment of software development and teacher training." Putting hardware into schools is a highly visible act. For politics, it seems enough to be able to say: "look what we are doing." This is even exhibited in the official plans, as Tucker states with respect to teacher training: "Official plans either give little consideration to, or ignore completely the effects of what they have proposed. There may be a pious paragraph about the need for good training, but in more than one instance the responsibility for providing that training is passed over to local colleges or teachers' organisations." Software and teacher training should obviously be the first concerns of anyone who thinks of the introduction of new technology.

1.7.4.2   *The introduction of NMIT may fail by lack of quality*

Here again, it is worthwhole to remember the problems with the audiovisual media. There are two kinds of poor quality which are relevant here.

(a)   Poor production quality

Programs were directed and edited by laboratory assistants, who had video as their hobby, university photographers who moved into the audiovisual field, and graduated from film schools who dropped out of the feature film business, and still tried to create their piece of art. Official training for the production of educational or instructional materials did not exist. As a result expensive products appeared which did not quite satisfy their objectives, and indeed showed many weaknesses as regards fitting into the curriculum.

(2)   Poor quality of the teachers' knowledge about media

Teachers never learned how to use audiovisual media, ad this is still the case, although - in the Netherlands - some initiatives in teacher training colleges may improve the situation (Schouwenaar, 1984; Wijnands, Van Outheusden, Veugelers & Lagarde, 1986).

Training is again the keyword, but not exclusively. The problem is that the necessity for training was not recognised. The mere fact that amateurs can easily take well-exposed photos or make beautiful super-8-mm films seemed to give the idea that the production of films and video programmes is easy. And the fact that it is in most cases rather obvious what is to be seen on a photo, gives the idea that the use of that photo in the classroom will be as easy as that. But: there exists no quality without effort. The production of a video programme is a professional job in which one tries to control all aspects of the communication problem as precisely as possible, preferably under studio circumstances. Control means effort. Similarly, the teacher has to integrate the use of that picture into the lessons by careful sorting out the possible function of the picture, and preparation of verbal guidance and desired student activities.

The message for NMIT is that effort is needed to identify the knowledge and skills necessary for the design, development and implementation of NMIT applications in education, distributed among the different roles of the persons concerned. New technologies require the attainment of new skills and the establishment of new co-operation patterns. If this is to be stimulated by the authorities, it demands an initial investment in training. The attainment of knowledge and skills may enhance the awareness of what quality is about, indeed acknowledging the roles of people from different disciplines with respect to the design, production and implementation of NMIT.

77

1.7.4.3   *The access to software is too difficult*

This again, is a recurrent problem. With respect to audiovisual media, teachers think it hard to locate appropriate materials. Audiovisual catelogues do not yield enough information to decide on the use of a program without first inspecting it. Viewing programs is time-consuming. And even then, two problems occur:

(1) Programs which are in principle usable, are often not used because the teachers do not see how to fit them into their lessons. Again, the lack of training in order to use media is most probably responsible. (Yet, some improvements are observed in recent years. Van Zon [1984] reports that the use of videotaped television lessons occurs more and more in a selective way. Secondary school teachers are more often using parts of programs, selected by themselves, and are more frequently explicitly integrating the programs into their lessons. Interestingly, this phenomenon is not caused by training. This issue will return near the end of this section.)

(2) Many programs are indeed hard to use. They are seldom designed as a component of an instructional method. Should this be the case, it would be easier to assess the possibilities to integrate them into the lessons.

With slight adaptations, these points count for NMIT as well. Especially with respect to CAL/CBT, many complaints about the lack of sufficient software packages are heard, with poor accessibility, questionable quality, and poor applicability as factors. Moreover, lack of standardisation is a serious problem. Tool-type software, like word-processing programs and spreadsheets, and databases are less vulnerable in this respect. Tucker (1987) observes that facilities of this kind seem now more and more to attract the novice computer user in the schools. Still, one has to learn how to use these facilities, while educational use has to be teacher-made in most cases, because of their general nature. Educational database management systems, item banking software, programs for keeping student records and test analysis, and packages for school management support take an intermediate position. In all cases, users must have the opportunity to develop experience, with training as the recommended initial step.

1.7.4.4   *The reward system to pay teachers is detrimental to the introduction of NMIT*

In the Dutch secondary school system, teachers are paid per lesson hour. They are supposed to be in the classroom with their students, and that is it. Activities outside the classroom, which are directly related to their lessons, are not separately remunerated. Lesson preparation and the correction of student work are done at home. Besides money for lesson hours, every school has some financial reserve for a limited amount of so-called "task hours". Task hours are used for activities of general interest, for instance, with respect to school management. For new activities, in most cases, hardly any extra time can be paid for. As a result, the effort for innovations, which in most cases goes far beyond the usual staff development activities and such, is nearly always an attack upon the teacher's own time. This may be alright with some enthusiastic computer pioneers, but it is unlikely that one can count on the idealism of teachers on a larger scale. Large scale innovations require temporary extra staff to get it off the ground.

Another thing is that the lesson hour concept seems to be sacred and 'division of labour' is almost unknown in the schools. People from the audio-visual field can report on the difficulties of introducing self-paced study systems or media-based self-study systems into the schools. With the students in a study-centre or in the library, the teacher has lost his class-room. How, then, can he or she be paid? Division of labour in terms of specialisation as curriculum developer, as mentor in self-study centres, as test expert, as courseware adviser, as classroom teacher and the like, seem hardly to be possible for the same reason. Tucker, who compared the integration of media into the curriculum in some fourteen countries, observes in this respect that in the Dutch secondary schools, in general, there is a rather traditional, teacher-dominated approach to learning. Whereas in other countries, alternative methodologies like individualised learning or open learning have been developed to the extent that media resource centres are installed at many places. Only few schools in the Netherlands have their resources organised in such a way (Tucker, 1986, p. 103). Although other reasons may prevail, it is again to be considered whether the payment per lesson hour reinforces the traditional teaching pattern. The lesson is that, for the time being, staffing consequences of NMIT have to take shape within the constraints of the currrent reward system. Possible changes in that system may, nevertheless, be object of further study.

1.7.4.5 *NMIT have to be cheap and easy to use*

Video discs, for example, have interesting features for use in schools, in particular for individual learning. But how will this medium ever have any impact if one single learning station costs about ten thousand guilders? It is out of the question that primary or secondary schools will be able to afford equipment of this price level in reasonable quantities. The only chance is that hardware, suitable for use in schools, will be available on the consumer market as well. Because of volume production, in that case, prices may appear to be low enough to accommodate the schools. This applies to any medium.

Ease of use forms the other factor. If it takes time to organise the use of the film projector (and the students have to move to a distant projection room to watch an educational film selected by the teacher), and then the control of the projector and the classroom illumination appears to be inefficient, the teacher will think twice before he or she decides to hire a film a second time. That cheapness and ease of use indeed form a factor, is supported by Van Zon (1984), who observed that the use of video in secondary schools started to increase significantly when cheap and easy-to-use video cassette recorders became available (81% of the schools started to use videorecorders after 1975). In his survey, teachers mentioned easy access to the equipment and ease of use of the equipment as important factors for the integration of video into the curriculum.

All NMIT are subject to these criteria:

- Hardware and software should be affordable. Some kind of mass production, therefore, seems to be necessary.

- Teachers should be in control of NMIT as a matter of course. The use of NMIT ought to be transparent to all users, teachers and learners alike. For the interaction with computer based systems, as Barbara Allen stated at ETIC-85 in London, this means that the user ought to have an answer to five questions at every possible moment: (1) Where am I?; (2) How did I get here?; (3) What can I do here?; (4) Where can I go to?; and (5) How is that accomplished?

In summary, the five points mentioned above are: (1) Hardware is never a problem. Software and teacher training are; (2) Quality applications mean effort on all levels by designers, producers and teachers. New technologies require new skills. Training may disseminate the available knowledge. Beyond this, time to gain experience should be available; (3) Easy access to software, which suits the needs of the teacher, is an important condition; (4) As they are paid for the lessons they give, teachers are obliged to work alone in their classrooms. This is a constraint to be dealt with; (5) Large scale implementation is only possible if the hardware-software combinations are cheap and easy to use.

These five points indicate that the implementation of NMIT in education needs a balance between:

(a) Skills and knowledge about NMIT
(b) Infrastructure (including hardware facilities
(c) The actual NMIT application, with software as a prominent component

The development of NMIT proceeds in phases. Research in laboratories yields results which are theoretically feasible for use in education. Next, several years of research and development are necessary to realise applications which are fit for wider use. From that point schools, and in particular the teachers in the schools, may start to incorporate the new possibilities into their curriculum. These three stages are schematised in Figure 2.

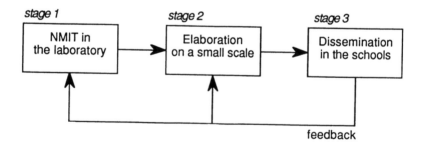

Figure 2: NMIT development

In relation to stage 1, technological advances are being translated into educational options. They may range from technical solutions without particular applications in mind, to courseware packages which are ready to teach learners how to attain certain learning objectives. This research and development is carried out by commercial producers and distributors or within teacher training colleges, universities, and government-paid institutions in the educational support system.

Sooner or later prototype solutions leave the laboratory, to be tested in the field in close co-operation with teachers. This is stage 2. In practice, stage 1 and stage 2 are closely related. Institutions which estimate the feasibility of a particular NMIT development in stage 1, continue to work on the promising ones in stage 2.

Stages 1 and 2 concern NMIT applications on a small scale. The dissemination of NMIT applications depends on what happens in stage 3. Here, it is obvious that the teachers are the ones who decide on the use of NMIT. Will they do so? According to Cerych and Jallade (1986) teachers need to develop a suitable attitude for adopting NMIT, apart from new knowledge and skills. This attitude includes:

"- openness to new developments and a readiness to assimilate new thinking,
- flexibility of approach to one's own teaching,
- readiness to collaborate with others,
- readiness to undertake re-training courses throughout a professional career,
- a view of the teacher role as one that combines the roles of innovator and researcher with a profound sense of professional responsibility." (Cerych and Jallade, 1986, p. 86).

The notable thing is, that this description seems to be valid for every teacher who takes his profession seriously. Why then is it presented as a change of attitude with respect to NMIT? Probably because the authors adhere to the philosophy that NMIT are indeed a blessing for the schools as soon as the teachers are at last willing to see this, while in fact the teachers are not ready to do so. Again, the resemblance with 'older' technologies, which are advocated in a similar way, indicates that one should be careful here. In this article, it is assumed that there are enough teachers in schools with a positive attitude towards the development of their profession, that on this point constraints are not to be taken into account.

The teacher is preparing his or her lessons, designing learning situations as a professional routine. NMIT applications can be a part of this routine. The main condition of success in stage 3 is that the teacher believes that the available NMIT options are fit for the intended purpose, and easy to use. Whether this condition is being fulfilled, depends upon the combination of the educational quality and user-friendliness of the technical facilities (hardware and software alike), and of the knowledge and skills of the teacher.

The balance between knowledge and skills, infrastructure and the eventual NMIT applications, has to be regarded in relation to the multidisciplinary nature of NMIT design, as was explained on the basis of Figure 1, with respect to the three stages of Figure 2.

The contributions from the different disciplines differ. Outside the schools, hardware and software problems should be solved, including the technical infrastructure. Applicable options may be designed, produced and marketed in close co-operation with the educational fields by people with audiovisual backgrounds, CAL/CBT backgrounds, or backgrounds with respect to hardware technology. If courseware production is at stake, curriculum specialists may pave the way for the introduction of particular NMIT applications in some subject matter area.

Inside the schools, applicable options have to be offered. This means that only proven techniques and software solutions should be made available. The teachers have to develop the pedagogical use of the new facilities. Technical problems should not bother them.

The balanced development mentioned earlier should maintain activities on two levels: (1) Research and development of educational applications of NMIT should keep in step with technical advancement. Here, stages 1 and 2 of Figure 2 are at stake. The exploration of research novelties and the development of new kinds of hardware and software take place on this level; (2) Implementation of new possibilities on a large scale. This is what should happen in stage 3. Here, didactic developments in the classroom are the central issue.

Laboratory research and classroom innovation are closely related. One has, however, to be aware of the time delay between the stages according to Figure 2. The goals of classroom innovation and of laboratory research should not be mixed up. Both need their own space to flourish.

The experiences with the audiovisual media show that somehow a balance will always develop. Education is anchored in a surrounding culture, and has its own culture at the same time. With some delay, schools adapt themselves to the demands and the way of life of the outside world. The use of educational television increased after the videocassette-recorder became available and the teachers became more familiar with television, probably not least because of viewing television at home.

Television programmes now take a modest but valued part of teaching-time, with history, world orientation, languages and geography as the subject areas where most applications are to be found (Van Zon, 1984). The use of other audiovisual programs, like tape-slide series or films, show a similar pattern, although the use of films is decreasing (Tucker, 1986). Teacher training with respect to audiovisual media is still rather limited. Tucker observes: "What appears to be missing ... is a significant number of courses which relate the use of media to particular methodologies. It may be that too many of the courses are still concerned with 'audiovisual aids' rather than with the integration and interaction of media and methods" (Tucker, 1986, p. 102-103).

Altogether, media are being used, but far less than the theoretical possibilities indicate. As soon as the need for alternative use of the media would develop, changes in this situation might be expected. Even without active policy on this point, changes will still be ongoing due to societal impulses, teacher initiatives and via incentives from academic media research. Such processes will eventually set trends. Active policy may speed up the development and make the direction of such trends more predictable.

An important difference between traditional audiovisual media and NMIT is that the necessity to pay attention to NMIT is felt more deeply, due to the observation that information technology (the 'IT' of NMIT) is affecting almost every branch of society, while the impact of audiovisual media has always been perceived to be more limited. Therefore, an increasing penetration of NMIT into the schools may be expected. Unguided implementation, as with audiovisual media, is unlikely to lead to optimal use of the potential of the new technologies.

As stated earlier, the three elements of balanced NMIT development are: (1) Knowledge and skills; (2) infrastructure and (3) development of applications. The lines along which NMIT innovations will take place, are represented in Figure 2.
With this in mind, the next conditions for successful implementation of NMIT may be formulated:

(a) Knowledge and skills should be taught. Teacher training should be aimed at integration of NMIT into the curriculum. All other participants from the different disciplines should be trained with respect to their desired contribution. Clear roles on the basis of Figures 1 and 2 should be defined to make this possible. This is a part of the infrastructure to be developed. 'Co-operation' is a second keyword next to 'training', to describe the crucial factors for the introduction of NMIT. Organised patterns for co-operation should be part of the infrastructure.

(b) The infrastructure should consist of several components:

- A structural organisational pattern with respect to Figure 2. Here, among others, a clear distinction should be made between teaching activities and activities with respect to systems design in general and courseware design in particular. This should emphasise what has to happen within or outside the schools.

- An infrastructure at school level to manage the NMIT resources, and to support teacher activities with respect to these media.

- An infrastructure at the macro level to control the balance by appropriate funding of the different components. Here, two princinples apply:

   (1) To guarantee continuous innovative power, investment in laboratory research is needed, to prevent the stream of technological advances from running dry;

   (2) The development of NMIT applications from the prototype stage to reliable components of curricula or of school management systems requires an amount of extra effort which is no part of the ordinary teacher's role. The macro level infrastructure should have provision for this extra effort. In the first place, this will be a matter of funding.

   Apart from this, a solution has to be found for the problems with respect to the lesson hour based salary-system in Dutch education.

(c) The use of NMIT applications is primarily a matter of curriculum development. On any level, NMIT ought to be approached accordingly. This is a principle that should guide the decisions with respect to training and infrastructure.

To meet these conditions, most probably a 'laissez faire' policy will not yield the required results. In fact, in most countries, government initiatives have been taken to stimulate the use of modern technologies in the schools. The question is whether the striving for a balance as described above, has been made sufficiently explicit and powerful to make the innovation oriented policy a success. If not, there will be a realistic chance that societal developments will overtake official policy. This may have a positive effect, but it may be as well that NMIT use stabilises on a relatively simple level, as is the case with audiovisual media. The trend towards word processing, the use of database packages, and the use of spreadsheets as the preferred modes of computer application in the schools should in this respect be observed with some anxiety. As the initial step of teachers to enter the NMIT world, this seems to be alright. There should, however, be a next step, which may only occur if the implementation of NMIT is carried out on the basis of an policy endorsed plan, which takes the above mentioned conditions into account.

The next section concludes this contribution with some remarks on the Dutch situation.

1.7.5   POLICY

In the Netherlands, several initiatives were taken by the government to stimulate the use of new technologies in the schools. With respect to NMIT, two have to be mentioned. The first one is the Information Technology Stimulation Plan (Informatica Stimuleringsplan or INSP), which started at the beginning of 1984. The INSP is organised in series of 'clusters', within which projects are carried out. The clusters are: Infrastructure (including the influence of information technology on the educational support system), first and second phases of secondary education, lower and middle vocational education, higher vocational education, adult and continuing education, in-service training for primary and special education, in-service training for secondary education, and research. The INSP has so far yielded a wealth of information about many aspects of the use of information technology in schools. Still, the large scale integration of information technology into the curriculum is not yet accomplished. In fact, the available information should be regarded as a starting point from which to develop a plan for the balanced development which was described in the last section. After the initial emphasis on hardware, software and courseware, it is now time to invest in time. What is needed is substantial amounts of time for teachers to be trained in the use of the new technologies in a more than casual way, and time to encourage serious curriculum development, with the new technologies within the schools. This time costs money. The focus of investments should no longer be with things to buy, but directed towards people to be paid. In the 'research cluster' of the INSP, the initiative has been taken to start two so-called experimental stations, consisting of a research group within a university and two secondary schools. These stations represent more or less stage 2 of Figure 2. This initiative may be a starting point to evaluate and to integrate the findings of the use of information technologies as they appear in the other INSP projects, and in the literature. It is too early to know whether this approach will be successful. At this point, it may be contended that it is at least very promising, provided that the teachers in the schools get enough extra time to participate actively in the developments of innovations. In general, however, analysis is needed to position the INSP attainments with respect to the three matters to be balanced, as they were described in the last section.

The second initiative to be mentioned is the installation of a steering committee for new media. This committee supports projects on all levels of education, which try to develop practical applications of the most current media. The emphasis is on interactive video disc and compact disc, viewdata systems and two-way cable communication. Research topics may concern: changes in the role of the teacher; consequences of the introduction of new media for the pedagogical situation; assessment of individual learning results, as well of the teaching situation; differences between target groups and school types with respect to the effects of the use of new media. The committee started its work in 1987. Results are not yet available. Here again, the impact of positive results of individual projects will be substantial if the committee succeeds in elaborating its policy in such a way that these results will fit into an overall structure which answers the three conditions of the last section.

In conclusion, training of the people involved, the development of patterns of co-operation between different disciplines, the acknowledgement of the development of NMIT applications, as a time-consuming job one has to pay for outside as well as inside the schools, active organisation of the necessary infrastructure, and re-emphasis of the attention towards curriculum development to integrate NMIT into the curriculum, are the relevant points in developing beneficial policy in favour of NMIT. The technology as such as at present not the problem. To ascertain future developments, however, it is necessary that the development of technical possibilities continues within the research laboratories
NMIT may be an advance for education. A balanced policy will decide when this will be noticeable.

1.7.6   BIBLIOGRAPHY

BAYARD-WHITE, C. (1985). An introduction to interactive video, London: Council for Educational Technology.

CERYCH, L. and JALLADE, J.P. (1986). The coming technological revolution in education. A report on the potential and limitations of new media and information technologies in education, Paris: European Institute of Education and Social Policy.

KLAUW, C.F. VAN DER, MEEUWEN, D.J. VAN and TIMMERMANS, L.J. (1987). Educatieve software voor het gebruik van multi-media banken; een inventarisatie. Rotterdam: Vakgroep Onderwijs Technologie, Erasmus Universiteit.

PALS, N. and VERHAGEN, P.W. (1987). DIDACDISC, Development and evaluation. Paper presented at ETIC-87, Southampton.

RUSHBY, N. (Ed.) (1987). Technology based learning, selected readings. London: Kogan Page.

SCHOUWENAAR, H. (1984). Audiovisuele media op de NLO's: een plaatje bij een praatje of toch iets meer ...?!, ID, Tijdschrift voor Lerarenopleiders, 5 (4), p. 196-205.

TUCKER, R.N. (1987). Information technology; the parson's egg of teacher training. Paper presented at ETIC-87, Southampton.

TUCKER, R.N. (Ed.) (1986). The integration of media into the curriculum. London: Kogan Page.

UNWIN, D. (1985). The cyclical nature of educational technology, Programmed Learning & Educational Technology, 22 (1), p. 65-67.

WIJNANDS, J., OUTHEUSDEN, J, VAN, VEUGELERS, H., and LAGARDE, W. (1986). AV-vorming: Knelpunten op de PABO. ID, Tijdschrift voor Lerarenopleiders, 8 (2), p. 77-82.

ZON, A.H.M. VAN (1984). Aanwezigheid en gebruik van televisie-apparatuur in het voortgezet onderwijs. Hilversum: Stichting Nederlandse Onderwijs Televisie.

1.8 COMPUTER-AIDED LEARNING: A SELF-DESTROYING PROPHECY?

by

Dr. JEF C.M.M. MOONEN,
University of Twente, The Netherlands

1.8.1   SUMMARY

It is becoming clear that the introduction of computers In education as a learning aid (CAL) has entered a critical phase. Demonstrable, measurable positive effects as a result of using CAL, and a general personal acceptance of computers will form the ultimate basis for success. To create those circumstances, computers will have to be introduced on a large scale. As a result of this effort, major changes will have to occur within the work habits of teachers, and in the organisational structure of each school. Those changes cannot be solved in a hurry, because aspects of a general innovational nature are largely involved. Every solution in this respect, therefore, needs time, much more time than was expected beforehand. Consequently, the speed of the introduction of computers in education, which has been very fast over the last 5 to 6 years, has significantly been slackened. No brute force nor great amounts of money can enforce a breakthrough. Only time and continuous training efforts can keep the ball rolling. All we can do now is to keep it rolling in the right direction.

The major obstacle for the integration of computers into the curriculum is the rigidity of existing organisational school and class structures. Because of the enormous numbers of teachers involved, their needs for in-service training, the discrepancy between the available amounts of hardware and what is needed and can be 'absorbed' by schools, a concentration upon a limited numbers of subject-areas seems to be appropriate, for instance, towards those curricula that are traditionally known as problem-areas (mathematics and mother-tongue). In those curricula much more flexibility will have to be created in order to give CAL the chance to prove its usefulness.

There is, in a general sense, enough evidence available to show that using CAL is effective, even cost-effective. However, one has to realise that as well as the medium, associated organisational aspects are equally responsible for the positive results. The intrinsic value of CAL programmes must therefore, be that through them, structural components of good educational practice implicitly have been, or will be, taken into account when using the programme in the classroom.

To create opportunities to face these problems, a four phase model for the development, research and implementation of educational software is presented, as well as the way this approach will be used in the Dutch situation.

1.8.2   THE THREE PRO'S: PROMISES, PROBLEMS AND PROSPECTS

It is becoming clear that the introduction of computers in education as a learning aid (for abbreviation purposes called 'computer-aided learning' or 'CAL') has entered a critical phase: success or failure will depend upon what will happen during the next two or three years. This situation did not come as a surprise. However, what causes its appearance?

A successful introduction of CAL is based upon four interrelated conditions:

- Availability of sufficient amounts of standardised hardware in schools;

- Availability of a large programme of pre-service and in-service training for teachers;

- Availability of sufficient amounts of good quality educational software;

- Integration of CAL into the organisation and the curriculum of the schools.

Of these, the first three conditions are necessary. However, they are not sufficient. It is the last condition which determines the impact of the previous ones.

Till now, major efforts have been concentrated upon the first two conditions. Although the connecting problems have not been completely solved yet, one can see that in many countries measures have been taken to come to large-scale introduction of standardised hardware in schools, and to set up training programmes for teachers to get them acquainted with the possibilities of the new technologies (Ennals, 1986; Plomp et al., 1987; Anderson, 1986; Winkler et al., 1986).

The problem of the production of educational software still remains unsolved. A solution in this respect will only be reached if an industrial-like production scheme is followed (Moonen, 1987). On the other hand, an industrial-like solution will only occur when there is a real market for the product. Nowadays, educational publishers have no confidence, or have lost it, in such a market (Van Dalen, 1985; De Vos et al., 1987). The educational software production problem is therefore turning around in a vicious circle. It will only come out of this, when national governments largely support such a production scheme, especially the research and development of it. In some countries there is a movement towards such a solution.

But what about the last condition: the integration of computers into the teaching-learning process? After the euphoria of the past years, it is now time for the promises to become reality. Will this be the case, or have we been fooling around all the time?

Why are we suddenly confronted with all the critical remarks? There are two answers to this question. It is clear that if one uses computers in the classroom for only a very limited amount of time, or in a very restricted situation, the positive impact on the work load and work habits of teachers and on the results and motivation of pupils is negligible. Contrarily, this kind of introduction will only cause trouble, because it will be experienced as an aberration of normal routines. Teachers who are enthusiastic about computers and education take these extra efforts for granted. But those teachers form only a small minority. The majority of them have to be convinced explicitly. It must be made absolutely clear that teachers and their pupils will gain in one way or another by using computers. As long as this evidence is not without any doubt, they will resist the introduction of computers, certainly because one also gets the impression that those computers will even threaten their jobs. Therefore, a better understanding of the possibilities, and restrictions, of computers by using them as much as possible, for instance, at home, will help to overcome their resistance.

Demonstrable measurable positive effects as a result of using CAL and a general personal acceptance of computers will form the basis for the last condition ever to have a chance for succeeding.
To create those circumstances, however, computers will have to be introduced on a large scale. As a result of this effort, major changes will have to occur within the work habits of teachers, and in the organisational structure of each school. Those changes cannot be solved in a hurry, because aspects of a general innovational nature are largely involved. Every solution in this respect, therefore, needs time, much more time than was expected beforehand.

Consequently the speed of the introduction of computers in education, which has been very fast over the last 5 to 6 years, has significantly been slackened. No brute force nor great amounts of money can enforce a breakthrough. Only time and continuous training efforts (in-service and preservice) can keep the ball rolling. All we can do now is to keep it rolling in the right direction.

In the following paragraphs some data will be presented in order to sustain these statements, and to present some evidence indicating which way to go further.

1.8.3 PATTERNS OF THE USE OF COMPUTERS IN EDUCATION

1.8.3.1 *Surveys*

There are data available about how computers are used in schools. Becker (1983, 1984, 1986) has published an enormous amount of detailed information about the situation in the U.S., investigated through a first survey between December 1982 and February 1983, and a second survey in the spring of 1985. The survey of 1985 included a sample of 2331 elementary and secondary schools, public and private.
In the Netherlands, the Ministry of Education Inspectorate (Ministerie van Onderwijs en Wetenschappen, 1986, 1987) conducted a survey at the end of 1985, in which all Dutch primary and secondary schools were included. Although the results of these surveys were published very recently, one should pay attention to the fact that the data themselves were gathered more than 18 months ago. This means that the current situation may differ significantly.

1.8.3.2 Number of computers

Becker reports that 'seven years ago, half of all high schools had no computer at all; and only four years ago, fewer than half of elementary schools had any'. In January 1983, 53 per cent of all schools in the U.S. had at least one microcomputer obtained for use in instruction (85% of all high schools, 42% of all elementary schools). In the spring of 1985 these percentages had risen to almost 100 and 80. A typical high school had more than 20 computers, a typical elementary school had 6.

In the Netherlands, the percentages at the end of 1985 were 91 for secondary schools and 28 for elementary schools. The expected numbers for 1986/87 are 95% and 59%. A typical secondary school had 9.7 computers, a typical elementary school had 1.6.
Because the educational system in the U.S. is fundamentally not different from the systems in other industrialised countries, and because Western Europe started approximately two years later with the introduction of computers in education, one can expect that the number of the U.S. will be met by West European countries within one or two years. Especially in elementary education a spectacular growth can be expected.

1.8.3.3 Use of computers

Becker states that 'a typical high school student could use computers to write compositions, memorize facts and vocabulary, understand relationships and concepts in mathematics and science and write computer programs'. He estimates that for each of these activities the student could use 30 minutes to three hours. This means as much as an hour or two per day, which translates to a student-computer ratio of 6 to 1 or 3 to 1. The numbers of the 1985 survey showed a ratio for high school students of 31 to 1, and for elementary school students a ratio of 60 to 1. These numbers illustrate clearly the discrepancy between an 'ideal' and the real situation.

1.8.3.4 Qualified teachers

Not only the amount of the hardware available determines the use of it. Teachers are also needed, to guide and stimulate the practical use of computers. Becker found out that 37% of elementary school teachers used computers regularly with students, versus only 15% of the secondary school teachers. Because secondary schools tend to be larger than elementary schools, the same number of teachers (5 per school) use computers regularly during the year. Of these teachers only 10% and 25%, respectively, could be considered as computer-knowledgeable, meaning that per two elementary schools, and for each secondary school one 'expert'-teacher is available.

1.8.3.5 Major instructional uses

The way computers are used by students is quite different in each school sector. Becker (1986) reports the distribution of computer activities in the figures quoted in Table 1.

Table 1: Distribution of computer activities (Becker, 1986)

Grade Span of School	Drill & Practice Tutorial	Discovery Learning Problem Solving	Pro- gram- ming	Word Proces- sing	Other	Total
K-6 Elem.	56%	17%	12%	9%	6%	100%
Middle/ Jr.High	30%	15%	32%	15%	9%	100%
High School	16%	10%	49%	20%	5%	100%

The numbers are percentages of the total instructional computer time in a school sector related to schools having computers.
In Dutch elementary schools, 17% of them use computers for CAL (drill and practice, tutorials, discovery learning, problem solving). Related to the schools who already have computers, this percentage rises to 61%. Thirty per cent of the lower secondary schools and 47% of the upper secondary schools use computers for CAL. Because most of these schools have computers, these percentages do not have to be adjusted. In 65% of Dutch lower secondary schools, computer literacy courses are given. In 60% of the upper secondary schools a course in informatics/programming can be taken.

In comparing these figures with those of the U.S., it is remarkable to see that especially in the elementary education of both countries, the use of CAL is widespread. In the Netherlands, a higher percentage for CAL appears in the upper secondary level, but this is due to the fact that a new mathematics curriculum has been introduced recently, in which computers and computing is one of the topics. Although the percentages for CAL suggest a widespread use of computers as a learning aid in Dutch education, a qualitative analysis reveals that these percentages do not mean that the impace of CAL is already of some significance.
Concerning informatics/programming in the upper secondary sector, the same tendency appears in the U.S. and the Netherlands. The same is true of computer literacy in lower secondary schools.

1.8.3.5.1   Regular or supplementary

Another approach can be made in distinguishing computer activities for remediation (help for students lagging in their understanding or achievement), for enrichment (activities apart from regular curriculum), and within regular instruction. The distribution is shown in Figure 1, as reported by Becker (1986).

As can be seen, in most cases the use of computers can be described as supplementary to regular instruction. Only in the highest classes of secondary school is the use of computers integrated in regular instruction.

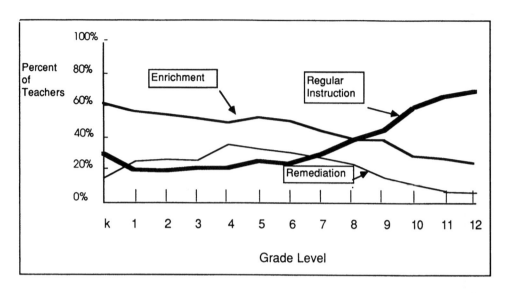

Figure 1: Distribution of computer instructional functions (Becker, 1986)

1.8.3.5.2   Major subject areas

'Mathematics and language arts (English and reading) are the major subjects for which computers are used in elementary schools. These two are joined by computer literacy as a third majoruse in middle schools. In high school, computer literacy and programming are the dominant subjects, with business education and mathematics following', says Becker (1986) in his report.

For the Netherlands, mathematics, mother-tongue and geography are the major subject areas in elementary school. In lower secondary schools major interest is towards computer literacy, mathematics and the physical sciences. The same is true in the upper secondary education with interest towards informatics/programming, mathematics, physical sciences and business education.

As can be seen, a remarkable similarity appears between the U.S. and the Netherlands with respect to this topic.

1.8.3.6   *Some patterns*

1.8.3.6.1   First observation

Large numbers of computers have found their way into education and with great speed, certainly in secondary education.

The introduction of large quantities of computers in schools has, in many cases, been stimulated by individual, private and uncoordinated efforts. As a result, the impact on the educational and organisational level has not unsufficiently been anticipated, especially in relation to the consequences on the innovational level.

1.8.3.6.2   Second observation

In spite of this spectacular growth, the amount of time a pupil can use a computer, is relation to the total school time available, is marginal.

This means that one cannot expect to see measurable effects of the use of computers on a significant scale yet.

1.8.3.6.3   Third observation

Only a minority of teachers is really involved in using computers in education, out of which only a very small proportion can be considered to be 'expert'.

Considering the number of teachers involved, the training of all of them in order to get them acquainted with computers and educational computing, will take a whole generation. There is no way to speed up this process. One has to realise that the efforts made now will have to continue for years to come.

1.8.3.6.4   Fourth observation

CAL is mostly used in elementary education in the area of mathematics and language teaching, mostly for drill and practice, tutorial purposes, and as a supplement to regular teaching.

The reason is most probably because of the appearance of a large number of problems in these subject-areas. In addition, the ease of use of drill and practice and tutorial programmes, and the great flexibility within the curriculum and the organisational structure of elementary education create opportunities for CAL.

1.8.3.6.5   Fifth observation

Use of computers integrated in regular instruction is mostly linked to classes in computer literacy or computer studies (informatics, programming).

As soon as the organisational structure creates possibilities for the use of computers, their usage becomes widespread.

1.8.3.7   Comments

The major obstacle for the integration of computers into the curriculum is the rigidity of existing organisational school and class structures. Because of the enormous numbers of teachers involved, their needs for in-service training, the discrepancy between the available amounts of hardware and what is needed and can be 'absorbed' by schools, a concentration upon a limited number of subject-areas seems to be appropriate, for instance, towards those curricula that are traditionally known as problem-areas (mathematics and mother-tongue). In those curricula much more flexibility will have to be created in order to give CAL the chance to prove its usefulness.

1.8.4   EXISTING EVIDENCE ABOUT THE IMPACT OF CAL

It is no longer sufficient to defend the use of CAL only on theoretical grounds. All of us know the marvellous stories about the potentialities computers have in education. As long as you are a 'believer', those arguments will be of value forever. If you have not joined this religion, you will ask for evidence. And there is evidence available now, to show that the use of computers has a positive impact on the quality and the productivity of education. It is even possible to show, under certain circumstance, that using computers is cost-effective when compared to other teaching-learning situations.

1.8.4.1   *Effectiveness*

Since the meta-analysis statistical method has been published by Glass et al. (1981), a great number of reviews about the effects of CAL have been published. Kulik and his colleagues started these publications in 1980, with reviews about effects of CAL in higher, secondary and primary education (Julik et al., 1980, 1983, 1985). Later, their studies were repeated by others, and augmented with new results. There was also some criticism about their publications: what was the influence on the results by unknown or uncontrolled variables, what about the quality of the teachers, the instructional material and the organisational structure involved, what about the amount of time effectively used within the instructional process? Another point of criticism has to do with comparing results connected with different kinds of CAL-applications.

Kulik and his colleagues accepted most of the criticism, and published new reviews in 1985 and 1986, using more stringent selection criteria for the studies involved (Bangert-Drowns, 1985; Kulik, 1987).
Recently, Niemiec and Walberg (1987) have examined a great deal of literature (a total of 16 reviews covering 250 different primary studies), in order to come to a comprehensive overview of the effects in different educational sectors. They conclude that 'the mean and median effect sizes are both 0.42, and the standard deviation is 0.08'. This means that the 'average and typical effect of CAL is to place the average student using it at the 665h percentile of traditional groups'.

More detailed analysis reveals the following conclusions. Concerning applications such as drill and practice programmes and tutorial programmes: these kinds of programmes have a definitely positive impact on the performance. The effect is greater with pupils below average and pupils in special education. The same is true in situations with a lower instructional level. Finally, there is a clear reduction in needed instructional time. The gain is minimal 20 to 30%.

Although conclusions for other kinds of applications are less clear, it seems appropriate to conclude that a positive effect of discovery learning (simulations, problem solving) is more likely to appear in higher education than in primary and secondary education. Many more details are available in other publications (Roblyer, 1985).

Cost-effectiveness

It is not surprising that we are able to measure improvement of results in education, when in a given situation a much greater amount of time, money, manpower, or other things have been invested than is normally available. A real appreciation of the results as mentioned above is, however, only possible if one includes in a comparison the use of CAL with traditional teaching methods, and the value of the extra investment needed in that particular situation.

Recently, Levin (1986) has compared the cost-effectiveness of four different interventions. He concludes that CAL is more cost-effective than other interventions, such as increasing the instructional time or reducing the class size. On the other hand, peer tutoring is more cost-effective than CAL.
Niemiec et al. (1986) have reacted to these results and explain that, based on their research, CAL is twice as cost-effective as peer tutoring (see Table 2).

Table 2: *Estimated effectiveness of four educational interventions in months of additional achievement gain per year of instruction for each $100 per student*

	Mathematics		Reading	
	Levin	Niemiec	Levin	Niemiec
CAL	1.0	4.5	1.9	3.5
Peer tutoring	4.6	2.9	2.2	1.0
Adult tutoring	0.8		0.5	
Increasing instructional time	0.5		1.2	
Reducing class size from to				
35 30	1.4		0.7	
35 25	1.2		0.6	
25 20	1.0		0.5	
35 20	1.1		0.6	

As well as a continuint discussion about the exact meaning and interpretation of this data and the use of the meta-analysis techniques (Slavin, 1986; Walberg, 1984), these data indicate that using CAL is at least as cost-effective as other kinds of interventions, possibly excluding peer tutoring.

On the other hand, CAL and peer tutoring can form a perfect match. Becker (1986) found that the largest reported impact of computers on instructional practices was by mutual assistance among students, meaning that a combined effort in this respect can optimise the effects. See also Johnson (1986). Supplementary to these results, Levin (1986) has investigated the use of a certain CAL-package in different U.S. school districts.

The results per district were quite different. The explanation Levin gave relates to the quality and quantity of the available personnel and implementation efforts per district. As well as the intrinsic quality of the educational software, the way it was used and supported highly influenced the ultimate results on the performance of the pupils.

These results are mostly based on experimental data related to drill and practice applications of CAL, for mathematics and reading in elementary education, and published in the Anglo-Saxon literature. It would be most interesting to have data available for other applications, other subject-areas, other school sectors, and from other cultures, for instance Eastern and Western Europe. One of the key-problems in relation to research about effectiveness and cost-effectiveness is the absence of relevant data, based on longitudinal and sound methodological methods.

1.8.4.3   Comments

The core of the criticism of effectiveness-studies is that the positive effect is not caused by the use of the computer, but that the use of different instructional methods, a different content, a different effort by the teachers and pupils, and the Hawthorne-effect produce the effects (Clark, 1985). Summarising, one could say that positive effects of using computers in education are not related to the medium, but to the circumstances the use of this medium forces us to create.
Now then, what is wrong with that? The main question is, if these better results ask for a more expensive instructional process compared to other approaches which lead to at least the same results. This is exactly what is expressed in cost-effectiveness ratios.

In addition, the study of Levin (1986) reveals that, as well as the medium, compelling surrounding organisational aspects (curriculum analysis and didactical analysis in the development phase, training and personnel efforts to support the implementation phase), are largely responsible for the positive results. The intrinsic value of CAL programmes, therefore, must be that through then, structural components of good educational practice implicitly have been taken into account, when using the programmes in the classroom. If these aspects create supplementary expenses, these costs have to be taken into account in the cost-effectiveness ratios.

1.8.5   RESEARCH, DEVELOPMENT AND IMPLEMENTATION

1.8.5.1   *Research and development in education*

Methodological issues concerning research and development in the social sciences are critical topics (Cooley, 1986). One could get the impression that research activities in the social sciences are sometimes hindered by a determination to reach a level of methodology that is comparable to that of the physical sciences. From a scientific point of view, this is certainly worthwhile. On the other hand, it is interesting to note that also in the physical sciences, research has often as its goal just to reach an optimal solution for an existing problem. The recent developments in the field of superconduction illustrate this very clearly. It is a fact that a comprehensive theory about human learning does not exist. This is also true with respect to the theoretical foundations for effective teaching methods. Considering the specific nature of humans, maybe these theories will never occur (Shulman, 1986). Because of this lack of didactical models, it is impossible to base the development of learning material, including computerised materials, directly upon theoretical grounds. As a consequence, a strategy based on craft and craftmanship has to be followed.

1.8.5.2   *Research, development and implementation in computer education*

Research about the use of computers in education is very complex, because of the very close connection with the development of the educational software, and the use of this software in a real classroom situation. Specifically in this case, the value of the research is highly dependent upon how relevantly variables and teaching strategies have been technically incorporated into a computer programme. The use of novelties in a real classroom situation is a common problem in the educational research area. It has, however, an extra dimension in this case, because a lot of technical aspects, causing a long tail of connected organisational problems, complicate the matter to a high degree. Therefore, there must be a very close relation between the development of educational software packages, the use of them for research purposes, the commercialisation of the product and the implementation. On the other hand, the expertise for making, investigating, distributing, implementing, and maintaining such packages must be sufficient, in order to be able to deliver software which makes optimal use of existing expertise in the different domains. To discuss this matter, the following distinction of activities is suggested.

1.8.5.3   *A four-phase model*

Development, research and implementation of computer activities in education requires a cycle of four distinguishable phases (see Figure 2)

First phase

The first phase starts from a real classroom situation, and tries to solve an existing teaching or content-oriented problem as well as possible, using the most appropriate technical possibilities. This phase includes formative evaluation (Moonen and Schoenmaker, 1986; Schoenmaker et al., 1987).

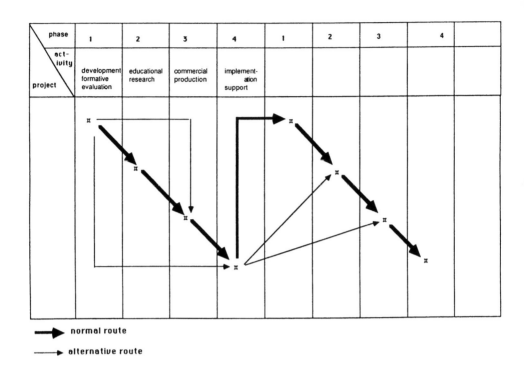

phase	1	2	3	4	1	2	3	4	
act-ivity project	development formative evaluation	educational research	commercial production	implement- ation support					

➤ normal route

→ alternative route

Figure 2: Development, research and implementation routes for educational software

In a general sense, a systems approach has to be followed here (Verhagen and Plomp, 1987). People working in this area can be characterised as 'educational technologists'. The result of their activity is a product which has the status of a 'prototype' and a 'protocol', which describes in a detailed fashion the consecutive activities, including the formative evaluation, in this phase.

Second phase

Further study of the protocols of the first phase will form the basis for 'pragmatic' theories which will only be valuable within a specific context, called 'shallow' theories by Suppes (1986), or 'miniature' theories by De

98

Klerk (1982). In order to enlarge the value of those theories, they will have to be examined vary carefully in experimental situations, in which sound methodological procedures can be followed. In the case of computers and education, studies about effectiveness and cost-effectiveness are essential. The results of those have to indicate the influence of the use of computers on the quality and productivity of education. There is a great need for this kind of research. One needs a clear picture in order to be able to defend the very radical changes needed as a result of a large scale introduction of computers in education.

This kind of work coincides with traditional educational research activities, and is traditionally performed by universities or connected institutes.

Third phase

In the third phase a final product is produced, based upon the knowledge gathered in the previous phases. This product has to take into account all the constraints that exist in real school situations: financial, technical, and organisational. One will also have to decide about the choice of the most appropriate instrumentation of the didactical process, in relation to the state of the art of the technology at that moment.

As far as industry is involved in the production process of learning materials, this third phase will have to be done by industrial companies: educational publishers or software houses.

Fourth phase

As soon as the commercial product is available, the pedagogigal implementation has to be carried out. Now all the known problems related to great-scale educational innovations appear: the need for information, for background material, for in-service and pre-service training, for extra facilities, and so on. This kind of support has to be given by teacher training institutes, educational support centres, and so on.

1.8.5.4   Comments

As well as the clear distinctions in different phases put forward in this model, the main message, however, is that different qualities are needed, in order to be able to optimise the results in each of the phases. In addition, and because of the great range of qualities concerned, not least on the technical level, those qualities will not be found within one kind of person. Consequently, different people have to carry out the different phases. Furthermore, those different phases will be characterised by a different 'working culture'. This approach causes new problems, especially on the managerial level, and the development of the interrelationships and connections of those phases. In a general sense, educational research in the era of information technology will have to shift away from a comprehensive individual approach towards a more industrial or technology-based approach. The lack of experience and tradition on the level of research management in this area has to be the major concern for the future.

In addition, the approach as described so far could bring about a solution for another severe problem. Recently, it became absolutely clear that educational publishers are very reluctant to enter the educational software market, or, if they have already done so, are pulling back. This attitude came about because of the discrepancy between the costs needed for research and development, and the profits that came out of selling the products. An approach in which the government takes the lead, and finances the prototyping and research aspects of educational software development, could solve this problem.

1.8.6   THE SITUATION IN THE NETHERLANDS

1.8.6.1   *The Advisory Committee on Education and Information Technology*

In 1982 the Advisory Committee on Education and Information Technology (AOI) proposed a strategy for the introduction of computers in education. In its first report, 'Information Technology: a necessity for everybody' (Plomp, 1985), the following recommendations were given:

1.   To introduce, within the next five years, computer literacy activities for all the pupils in the age range of 12-15.

2.   At the same time, to stimulate the use of computers in schools to support administrative and organisational aspects directly related to educational aspects.

3.   In the medium term, to stimulate preparations for the introduction of computers as a learning aid, especially as a tool.

4.   In the long term, to start thinking about using computers as a replacement for teachers.

This report has had a definite influence on the Dutch policy since that time (Moonen, 1986).

1.8.6.2   *The Informatics Stimulation Plan*

The main activities in the Netherlands at present in the field of computers and education are concentrated in the educational part of the Informatics Stimulation Plan (INSP). The INSP is a five-year programme, 1984-1989, to stimulate the use of computers in all the educational sectors, excluding university teaching, with a budget of over 280 million guilders for the educational sector (Van Deursen, 1987).

Within the INSP, special attention has been given towards the creation of an infrastructure (including a methodology) for educational software development; the development and implementation of computer literacy courses and software packages in the different school sectors, with a clear (financial) priority towards vocational education, and a clear reserve towards elementary education; the organisation of large programmes for pre-service and in-service training; the stimulation of research.

Now, five years after the AOI-advice, we can see that the introduction of computer literacy is going well in Dutch education. And although the stimulation of the use of computers to support administrative and organisational activities has not received major attention within the current policy framework, these developments are also going well. Approximately 55% of the secondary schools and 15% of the elementary schools that have computers, use them in this respect. The use of computers in vocational education has got a major impulse, especially by enlarging the budget for the inventory of the schools. Available in-service courses are constantly booked up (PSOI, 1987). Because of a limited budget, and because of disagreement about the main trends to be investigated, the research efforts till now have not been too successful.

The Netherlands now faces the challenge to shift towards execution of the third recommendation of the AOI: the introduction of computers as a tool in education. In paragraph 1.8.3 we have already indicated that this new road is full of obstinate difficulties. A major concern in this respect is how to obtain sufficient educational software available in order to be able to build up educational packages which enable a 'continuing and frequent use of it in education through which a measurable and positive impact on present education processes can be realised' (Kooreman and Moonen, 1987).
In the previous paragraph an industrial-based approach was formulated to create a context in which this problem could be tackled.

1.8.7   FUTURE STRATEGIES

Recently, the Ministry of Education and Science of the Netherlands has started two new projects in order to give a concrete interpretation to activities in this direction: a courseware development project called 'POCO' (courseware development for computers in education), and a pilot-project for educational research concerning, and implementation of, computers in schools, called the 'Experimental Stations'-project. POCO is related to the first and third phase of the model of paragraph 1.8.5, the Experimental Stations-project is connected to the second and fourth phase of this model (see Figure 3).

1.8.7.1   *POCO-project*

The POCO-project will reorganise a number of existing activities in the field of educational software development, and concentrate the efforts more specifically towards the development of sufficient amounts of software for a restricted number of subject-areas within the elementary and secondary school sectors, including lower and middle vocational education. The project will start at the end of 1987 and will have a life-span of four years. The Centre for Education and Information Technology (COI) will manage this project. The main purpose of the project is to produce, within fixed time limits, a reasonable amount of good quality educational software, in order to create the basis for frequent use of it within certain subject-areas, which can significantly influence the outcomes, the motivation and the working habits of teachers and students.

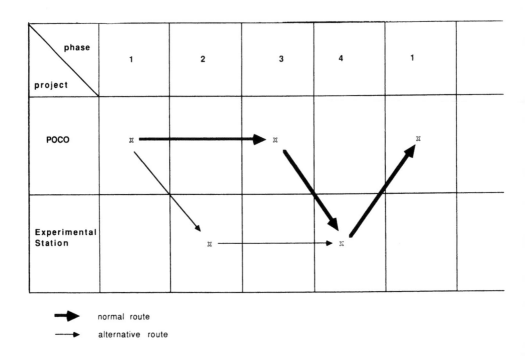

Figure 3: Connections between POCO and the Experimental Station-projects and the routes of educational software development

The project will be of a cyclical nature. Each of the cycles will consist of four phases:

-   Choosing the subject-areas, and within those areas, specific topics, for which use of computer can be of any help. Setting up priorities in relation to the financial constraints and political arguments.

-   Production of so-called product descriptions, in which at least a functional description of the programme wanted has to become available. In addition, and in a more general fashion, standardised procedures concerning the project and development methodology and documentation will have to be taken into account.

In both phases, the views of practitioners and the educational support system of the Netherlands will have to be organised.

- Via a tender procedure the technical production of the programmes will be contracted out.

- The distribution of the produced packages will be organised through the normal channels of the educational publishers. These publishers will be requested to engage in specific areas of the whole process.

For the POCO-project a total budget of 26 million guilders is available.

1.8.7.2 *Experimental Stations-project*

In 1986 the decision was taken to start a project with two so-called 'Experimental Stations'. An 'Experimental Station' is a unit of two pilot schools in the general secondary sector, combined with a group of educational researchers, connected to a Dutch university. It is the intention to make available within each of these pilot schools, a level and amount of hardware and (educational) software that can only be expected in the majority of Dutch schools, after a period of 5 to 10 years from now. It is the task of the educational research group to investigate questions, in a concentrated and interrelated way, connected with the implementation of computers in the real classroom situation.

Two experimental stations now exist: one in the western part of the Netherlands, connected to the Free University of Amsterdam and the State University of Utrecht, and one in the eastern part of the country, connected to the University of Twente

The project will be worked out in two phases:

- The first and starting phase from May 1987 till July 1988.
- The second phase from July 1988 till July 1992.

For the first phase of the Experimental Station-project a total budget of 2.3 million guilders is proposed.

1.8.8 CONCLUSION

As is shown in Figure 3, these two projects must make it possible in the Netherlands to face the problems as described in Section 1.8.3. The political decisions have been taken, and the financial means are available. The major problem still to be solved is the organisation of a smooth-running project management. What a challenge!

1.8.9   BIBLIOGRAPHY

Section 1.8.2

ENNALS, R., GWYN, R., ZDRAVCHEV, L. (Eds.) (1986), Information Technology and Education: The Changing School. Chichester: Ellis Horwood Publications.

PLOMP, Tj., DEURSEN, K. VAN, and MOONEN, J. (Eds.) (1987), CAL for Europe: Computer-aided learning for Europe. Amsterdam, North-Holland.

ANDERSON, R.E. (Ed.) (1986), National Educational Computer Policy Alternatives. New York: Association for Computing Machinery.

WINKLER, J.D., STASZ, C., and SHAVELSON, R. (1986), Administrative Policies for Increasing the Use of Microcomputers in Instruction. Santa Monica: The Rand Corporation.

MOONEN, J. (1987), Educational Software Development: the Pedagogical Design. In: PLOMP, Tj., VAN DEURSEN, K., and MOONEN, J. (Eds.), CAL for Europe: Computer-aided learning for Europe. Amsterdam, North-Holland.

DALEN, J. VAN (1987), The Courseware Dilemma. In: MOONEN, J., and PLOMP, Tj. (Eds.), EURIT86, Developments in Educational Software and Courseware. Oxford: Pergamon Press.

DE VOS, J.A.M., and BENDERS, F.T. (1987), Courseware voor het voetlicht [Courseware put on the stage]. Utrecht: Groep Educatieve Uitgeverijen (in Dutch).

Section 1.8.3

BECKER, H.J. (1986), Instructional Uses of School Computers, No. 1, 2 and 3. Center for Social Organization of Schools. Baltimore: The Johns Hopkins University.

BECKER, H.J. (1983, 1984). School Uses of Microcomputers, No. 1, 2, 3, 4, 5 and 6. Center for Social Organization of Schools. Baltimore: The Johns Hopkins University.

Ministerie van Onderwijs en Wetenschappen, Onderwijs en Informatietechnologie, Inspectierapport 9 en 12 (1986, 1987). Den Haag: Staatsuitgeverij (in Dutch).

Section 1.8.4

GLASS, G., McGAW, B., and SMITH, M. (1981), Meta-analysis in Social Research. Beverly Hills, Cal.: Sage Publications.

KULIK, J., KULIK, C., and COHEN, P. (1980), Effectiveness of computer-based college-teaching: a meta-analysis of findings. Review of Educational Research, 50, 4.

KULIK, J., BANGERT, R., and WILLIAMS, G. (1983). Effects of computer-based teaching on secondary school students. Journal of Educational Psychology, 71, 1.

KULIK, J., KULIK, C., and BANGERT-DROWNS, R. (1985). Effectiveness of Computer-Based Education in Elementary Schools. Computers in Human Behavior, 1.

BANGERT-DROWNS, R., KULIK, J., and KULIK, C. (1985). Effectiveness · of Computer-Based Education in Secondary Schools. Journal of Computer-Based Instruction, 12, 3.

KULIK, C., and KULIK, J. (1987), Effectiveness of Computer-Based Education in Colleges. AEDS Journal, in press.

NIEMIEC, R., and WALBERG, H. (1987), Comparative effects of computer-assisted instruction: a synthesis of reviews. Journal of Educational Computing Research, 3, 1.

ROBLYER, M. (1985). Measuring the impact of computers in instruction: a non-technical review of research for educators. Washington: Association for Educational Data Systems.

LEVIN, H. (1986), Cost and Cost-Effectiveness of Computer-Assisted Instruction. In: CULBERTSON, J., and CUNNINGHAM, L. (Eds.), Microcomputers and education: 58th Yearbook of the NSEE. Chicago: The University of Chicago Press.

LEVIN, H., and MEISTER, G. (1986). Is CAI Cost-Effective? Phi Delta Kappan, 67, 10.

NIEMIEC, R., BLACKWELL, M., and WALBERG, H. (1986), CAI can be doubly effective. Phi Delta Kappan, 67, 10.

SLAVIN, R. (1986). Best-Evidence Synthesis: an alternative to meta-analytic and traditional reviews. Educational Researcher, 15, 9.

WALBERG, H. (May, 1984), Improving the Productivity of America's Schools. Educational Leadership.

JOHNSON, R., JOHNSON, D., and STANNE, M. (1986). Comparison of Computer-Assisted Cooperative, Competitive and Individualistic Learning. American Educational Research Journal, 23, 3.

CLARK, R. (1985), Confounding in Educational Computing Research. Journal of Educational Computing Research, 1, 2.

LEVIN, H. (1986). Cost-Effectiveness of Computer-Assisted Instruction: Some insights. To be published in the Proceedings of the International Conference on Courseware Design and Evaluation. Israel Association for Computers in Education.

Section 1.8.5

COOLEY, W., and BICKEL, W. (1986). Decision-Oriented Educational Research. Boston: Kluwer-Nijhoff Publishing.

SHULMAN, L.S., and RINGSTAFF, C. (1986). Current Research in the Psychology of Learning and Teaching. In: WEINSTOCK, H., and BORK, A., Designing Computer-Based Learning Materials. NATO ASI Series. Berlin: Springer Verlag.

MOONEN, J., and SCHOENMAKER, J. (1986), Production Techniques for Computer-based Learning Material. Paper presented at the annual AERA meeting. San Francisco: Eric Clearinghouse on Information Resources, ED 276 406.

SCHOENMAKER, J., MAST, C. VAN DER, and MOONEN, J. (1987). A methodology for the Development of Educational Software. In: MOONEN, J., and PLOMP, Tj. (Eds.), EURIT86, Developments in Educational Software and Courseware. Oxford: Pergamon Press.

VERHAGEN, P., and PLOMP, Tj. (1987). Educational Technology: a Dutch contribution to the debate. Paper presented at ETIC 87, Southampton, 15th April. Enschede: University of Twente, Department of Education.

SUPPES, P. (1986). Paper presented at the annual AERA meeting. San Francisco.

DE KLERK, L.F.W. (1982), Uitdaging en Teleurstelling. In: VAN DER KAMP, L., and VAN DER KAMP, M. (Eds.), Methodologie van Onderwijsresearch [Methodology of Educational Research]. Lisse: Swets & Zeitlinger (in Dutch).

Section 1.8.6

PLOMP, Tj., and MUYLWIJK, B. VAN (1985). Information Technology in Education: Plans and Policies in the Netherlands. In: DUNCAN, K., and HARRIS, D., Computers in Education, Proceedings of the 4th World Conference on Computers in Education. Amsterdam, North Holland.

MOONEN, J. (1986), Toepassing van computer-systemen in het Onderwijs [Application of computer systems in education]. Wetenschappelijke Raad voor het Regeringsbeleid, Voorstudies en Achtergronden, V52. Den Haag: Staatsuitgeverij (in Dutch).

DEURSEN, K. VAN (1987), The Introduction of Information Technology in the Dutch Educational System. In: PLOMP, Tj., VAN DEURSEN, K., and MOONEN, J. (Eds.), CAL for Europe: Computer-aided learning for Europe. Amsterdam, North-Holland

PSOI, Computers in Education: a future-oriented analysis (1987). PSOI-reeks nr. 22. Den Haag: Staatsuitgeverij.

KOOREMAN, H.J., and MOONEN, J. (1987). Beleidsonderzoek Proefstations. Inventarisatie van Praktijkervaringen en Aanknopingspunten voor verder Beleid. In: PSOI-reeks nr. 25, Ministerie van Onderwijs en Wetenschappen. Den Haag: Staatsuitgeverij.

PART 2:
NATIONAL REPORTS AND BACKGROUND INFORMATION

2.1 INFORMATION IN BRIEF ON DOCUMENTATION AVAILABLE

Some information sent to the organisers of the Workshop on Interactive Learning and New Technologies was too lengthy for inclusion in this Reader. Here follows a short (bibliographic) description of this information. To facilitate contacting persons and instructions, addresses have been added, where possible.

2.1.1 INFORMATION FROM FINLAND

Integration of information technology in school education (the "TOP-project"). Helsinki, National Board of Vocational Education, the National Board of General Education, 1986. (10 p.)

2.1.2 INFORMATION FROM FRANCE

Two brochures from the "Association régionale pour le développement de l'enseignement multimédia informatisé" (ARDEMI), 93 Chemin des Mouilles/ B.P. 167, 69130 Ecully, France, 1985. (3 p. each).

A leaflet announcing the existence of the organisation "Réseau Jeunes et Technologies", Tour Periféric, 6 Avenue R. Reynaud, 93306 Aubervilliers.

A poster from CESTA (Centre d'Etudes des systèmes et des Technologies Avancées), 1, Rue Descartes, 75505 Paris.

Durey, A. and R. Journeaux, Apport des recherches en didactique dans le développement futur de l'utilisation des ordinateurs dans l'enseignement des sciences physiques. A leaflet from the L.I.R.E.S.P.T. (Laboratoire Interuniversitaire de Recherche sur l'enseignement des sciences physiques et de la technologie, Tour 2B, 5e étage, couloir 2313, 2 Place Jussieu, 75251 Paris. (5 p.).

From the World Federation of Teachers' Union (Fédération internationale syndicale de l'enseignement) some articles were sent in. They were copied from two sources: Le travailleur de l'enseignement technique, journal destiné à nos syndiqués, and l'IRETEP, revue de l'institut des enseignements techniques et professionels.

The articles copied are:

Informatique en LEP: oui, d'accord, mais pas n'importe comment (Le travailleur de l'ET, no. 336, 1982);

Expériences pédagogiques dans les LEP du tertiaire (Le travailleur, no. 340, 1983).

Avoir un enseignement polytechnique dans les collèges introduire de véritables enseignements de caractère technologique (id.).

25-11 Mécanique générale. Extrait de l'intervention de la Fédération de métallurgie (Le travailleur, no. 346, 1984).

Stein, Evelyne. L'ordinateur: pas instrument miracle mais instrument ressource (IRETEP, no. 2).

Huynh, Jeanne-Antide. Défense et illustration de l'ordinateur à l'école (id.).

Collinot, R. La micro-informatique au service de l'enseignement du français dans les lycées professionels (id.).

Rigaud, Françoise. Du traitement des élèves au traitement de texte (id.).

Effets du micro-ordinateur sur le rapport à l'écrit des élèves de LEP (IRETEP, no. 6).

Cru, Didier. Recherche et dessin Technique (IRETEP, no. 7).

2.1.3   INFORMATION FROM ITALY

A brochure drawing attention to the existence of SASIAM, The School for Advanced Studies in Industrial and Applied Mathematics, Teconopolis, Strada Prov. le per Casamassima, km. 3, I-70010 Valenzano (Bari), Italy.

2.1.4   INFORMATION FROM THE UNITED KINGDOM

A leaflet drawing attention to the Journal of Computer Assisted Learning. Oxford (etc.), Blackwell Scientific Publications. ISSN: 0266-4909.

CET and IT in education and training. London, Council for Educational Technology for the United Kingdom (March, 1987). Address of CET: 3 Devonshire Street, London W1N 2BA.

Investigations on teaching with microcomputers as an aid; a catalogue of resources by members of the ITMA-Collaboration and collaborating groups and individuals. (15 p.) The principal bases of the ITMA-collaboration are: College of St. Mark and St. John, Derriford Road, Plymouth PL8 8BH, and ITMA-Collaboration, Shell Centre for Mathematical Education, University of Nottingham, Nottingham NB7 2RD.

Interactive video register. Glasgow, SCET. Address of SCET: 74 Victoria Crescent Road, Glasgow G12 9JN.

2.2 LIST OF ONGOING OR COMPLETED RESEARCH

compiled by
the Section for Educational Research
and Documentation, Council of Europe

DENMARK

Prof. Jorge Aage JENSEN
Institute of Informatics
Royal Danish School of Educational Studies
115 B Emdrupvej
DK-2400 COPENHAGEN NV

The main efforts have been concentrated upon the adaptation of a Danish version of a knowledge-based system shell (DLH-MITSI), for use in schools (seventh through eleventh grade). The adaptation builds on a system developed in the United Kingdom by Jonathan Briggs, called MITSI (Man In The Street Interface), which is an introduction to essential features of logic programming (the system is written in LPA Prolog Professional), as well as a system shell for building knowledge based systems ("expert systems").
Work in the schools will concentrate upon one aspect of the system: a tool for analysing knowledge. The main assumption is that it is pedagogically and psychologically important to have a tool available, by which the issued of knowledge construction, knowledge representation and knowledge acquisition and use may be made "concrete". Analysis of a selected domain may be undertaken by pupils, and teachers and pupils, cast in the shape of constructing a DLHMITSI-programme. Such a programme consists of an information base, coupled with one or more rule-bases drawing on the inference-engine built-in in the underlying Prolog, on the basis of a conceptual analysis of the domain.

The system is intended to be used both within a particular subject, e.g. geography, and cross-curricula-wise, e.g. establishing a knowledge base about nutrition.

The work is part of an international group effort, under the name of "Prolog in Education Group (PEG)", originating at the University of Exeter (Jon Nichol).

FINLAND

NATIONAL BOARD OF EDUCATION
Research & Development Bureau
Hakaniemenkatu 2
SF-00530 HELSINKI

1. Development of a syllabus for <u>computer studies in upper secondary education</u>. The aim is to find useful course material for a nationwide revision of syllabuses in computer studies.

2. Development and testing of <u>computer-assisted teaching programmes</u> for various subjects in <u>upper secondary education</u>; these programmes will be linked to the curricula. Sixteen schools are involved.

3. Testing of <u>computer programmes in primary education</u> (lower forms of the comprehensive school). The aim is to develop programmes which help to familiarise children with the world and the use of computers.

4. Development of a syllabus for <u>computer studies</u> as an optional subject in <u>lower secondary education</u> (upper forms of the comprehensive school).

5. Development and testing of <u>computer-assisted teaching programmes</u> for various subjects in <u>lower secondary education</u>; assistance to schools developing their own software.

FEDERAL REPUBLIC OF GERMANY

BUNDESMINISTERIUM FÜR BILDUNG UND WISSENSCHAFT
Postfach 20 01 08
D-5300 BONN 2

Financial support for numerous research and pilot projects concerning the NIT and education. In nearly all these projects interactive learning as a result of introducing NIT plays an important role. The full list of the projects (document BMBW-I B5/II B4 of 17 September 1986) may be obtained from the Ministry.

M. LANG, J. LEHMANN, D. SINHART
Institut für die Pädagogik der Naturwissenschaften an der Universität Kiel
IPN - Gebäude N30, Olshausenstr. 40
D-2300 KIEL 1

Research into the attitudes of secondary school pupils regarding computer uses and their experience with computers.

VEREIN ZUR FÖRDERUNG DER PÄDAGOGIE DER INFORMATIONSTECHNO-
LOGIEN
Südstrasse 135
D-5300 BONN 2

Development of teaching and learning material

1. Development of models of retraining unemployed teachers to become specialists in computer education.

2. Projects: NIT - computer camps, seminars (Neue Technologien, computer-camps/seminar)

    Pilot experiments with the organisation of two computer holiday camps to familiarise with NIT:

    -   one for the 12-16 age group, and
    -   one for the 16-25 age group.

3. Project "Use of NIT in youth centres" (Der Einsatz von neuen Techno-logien in Häusern der offenen Jugendarbeit).

    Aim: to promote interactive learning by means of computer games in leisure-time activities.

4. Project: NIT computer clubs

    Organisation of Technology Computer Clubs (TCC), e.g. one in Mosel-strasse 7, D-55000 TRIER. The clubs offer computer programmes for interactive learning to schoolchildren, adolescents and adults, as well as general introduction into the NIT.

5. Project: "YOUTH-COMPUTER-CONSULT" to train staff engaged in youth activities for the use of NIT in youth programmes ("Senioren-Animation" gegen "Jugendarbeitslosigkeit")

6. Research into NIT and interactive learning in the case of pupils (spe-cial education) and adolescents facing learning difficulties (Neue Tech-nologien in der An- und Verwendung bei lernbehinderten Jugendlichen im Übergang von der Schule in die Berufspraxis).

    Pilot project supported by the Federal Ministry of Education in Bonn, the Schleswig-Holstein Ministry of Education and others.

7. Research into the use of NIT for improving language learning in the case of young second generation migrants.

    Development of interactive computer software for seminars meant to familiarise adolescents with problems of economy and environment. (Sprache und Computer bei jungen Ausländern der 2. Generation.)

GREECE

Dr. Emmanuil PAPAMASTORAKIS
University of Creta
Department of Mathematics
PO Box 470
GR-IRAKLION/Creta
(also at TU Berlin, Fachbereich Mathematik, Strasse des 17. Juni 135,
D-1000 BERLIN (WEST) 12)

Research into computer studies at school with stress on in-service education
and training of teachers (IBSET).

ITALY

Rosa Maria BOTTINO
Istituto Per La Matematica Applicata
Consiglio Nazionale Delle Richerche
Via LB Albert 4
I-16132 GENOVA

Responsible for programming a new mathematics project.

NETHERLANDS

Joost BREUKER, Bert CAMSTRA, Stefano CERRI, Peter MATTIJSSEN,
Markus VAN DIJK
Centre for Research into Higher Education
Oude Turfmarkt 149
NL-1012 GC AMSTERDAM

Intelligent computer-assisted instruction: development of the DART authoring system and research on misconceptions and foreign language learning
(Project No. 3430 in the EUDISED Data Base).

G. ERKENS
Teaching and Research Group on Education
State University of Utrecht
Heidelberglaan 8
NL-3508 TC UTRECHT

Analysis of dialogues in interactive problem-solving (continuation of project
described under the name of G. KANSELAAR).

Research questions:

1.   What is the relationship between cognitive and communicative processes
     during interactive problem-solving?

2. How can this knowledge be incorporated in an "intelligent" dialogue monitor for computer-assisted instruction?

W.N.H. JANSWEIJER, J.J. ELSHOUT, B.J. WIELINGA, D.J. BIERMAN, J.A. BREUKER
Institute for Cognition Research
Weesperplein 8
NL-1018 XA AMSTERDAM

Development of a computer coach for thermodynamics (physics) (Project No. 5004 in the EUDISED Data Base).

G. KANSELAAR
Teaching and Research Group on Education
State University of Utrecht
Heidelberglaan 8
NL-3508 TC UTRECHT

Analysis of dialogues in interactive problem-solving

In order to realise a "natural" and coordinated dialogue between student and tutor in computer-assisted instruction, more knowledge about the relationship between cognitive and communicative processes during interactive problem-solving is needed. For this purpose natural task dialogues between pupils (aged 10-12) will be analysed. The goal of analyses is to build models, which are to stimulate the way in which cognitive information processing influences the structuring and use of dialogue-acts during co-operation. The resulting models of simulation will be used to construct an artificial intelligence dialogue monitor. The monitor will be implemented in a computer-assisted task of problem-solving, and will operate on a student-based (novice) system. Effects on students' problem-solving of this task will be determined for experimental conditions of monitor control.

SWEDEN

Berner LINDSTRÖM
Gothenburg University
Department of Education and Educational Research
Box 1010
S-43126 MÖLNDAL

1. Interactive didactical environments (IDM)

    Aims: -   the exploration of the computer as a component in a teaching-learning system aimed at promoting conceptual change;
            -   humans learning and thinking computer use.

2. Computers in the learning and teaching of reading and writing.

117

Jean PASCHOUD
Centre vaudois de recherches pédagogiques (CVRP)
Martery 56
CH-1005 LAUSANNE

Observations of an experiment with computers: a study of activities of a class with pupils in the eighth grade.

(Study published by Swiss Co-ordination Centre for Research in Education, Francke-Gut, Entfelderstrasse 61, CH-5000 AARAU).

UNITED KINGDOM

Derek BALL, Barry GALPIN
School of Education
University of Leicester
21 University Road
GB-LEICESTER LE1 7RH

SPIRAL Project (an interesting attempt to use a PROLOG shell with children).

Jonathan BRIGGS
Information Technology Development Unit
Kingston College of Further Education
Kingston Hall Road
Kingston Upon Thames
GB-SURREY KT1 2AQ

Work on an expert system project in further education. Production of software.

Judith CHRISTIAN-CARTER
Project Manager
Council for Educational Technology
3 Devonshire Street
GB-LONDON W1N 2BA
(Tel. 01 580 7555 or 01 636 4186)

Tom CONLON
Moray House College of Education
Holyrood Road
GB-EDINBURGH EH8 8AQ

1. Development of an expert system for teacher training.

2. Engaged in two projects:
    -   MAPOL (Machine Assisted Planning of Lessons) - an application of expert systems to education;
    -   Logic Programming and Education - the use of PROLOG and PAR-LOG as problem-solving tools for learners and as implementation technologies for AI-CAL (Artificial Intelligence - Computer-Assisted Learning)

COUNCIL FOR EDUCATIONAL TECHNOLOGY FOR THE UNITED KINGDOM
3 Devonshire Street
GB-LONDON W1N 2BA

Educational applications of information technology: a UK database of research.

Professor R.E. LEWIS
Economic and Social Science Research Council
Information Technology and Education Programme
Department of Psychology
University of Lancaster
GB-LANCASTER LA1 4YF

Responsible for the ESRC IT (Information Technology) Education Programme and an Expert Systems/Teacher Training Project.

MORAY HOUSE COLLEGE OF EDUCATION
Holyrood Road
GB-EDINBURGH EH8 8AQ

Responsible research worker: Mr. Tony VAN DER KUYL

Interactive Videodisc Project (development programme project in environmental studies in primary education).

In December 1985 the Department of Trade and Industry announced a two-year one million pound programme to support and investigate Interactive Videodisc production, and its evaluation in the United Kingdom education system. Moray House is responsible for the production of one of the eight discs to be produced by this United Kingdom project. The college had in 1984 mounted its own feasibility study of this new educational applications technology, and this work provided a basis for its contribution to this national initiative.

The Moray House project is also an integral element of the Primary Education Development Programme (PEDP). The curriculum focus, environmental studies, is aimed at both teachers and learners, and duel audio channels and external software should ensure versatility of use. The structure of the videodisc will facilitate teacher decision-making in areas such as resource management, information retrieval procedures, methodology, and also cover different approaches of exploring/developing primary projects in this area. The ability of the videodisc also to establish the dynamic relationship between classroom action and fieldwork, offers a unique opportunity for the teacher's staff development.

PROLOG Education Group (PEG)
Exeter School of Education
St. Luke's
GB-EXETER EX1 2LI

PEG is running three intermeshed projects:

1. Primary schools project based on three feeder primary schools to our secondary school. This is a pilot scheme to integrate the software within the primary school curriculum.

2. Specific Learning Difficulties - dyslexic pupils. A project which applies the software to the learning of SLD pupils in both primary and secondary schools.

3. Secondary schools - a project for the use of the shells with Humanities pupils, and in the teaching of Latin.

4. A curriculum development project: PROLOG in the classroom (responsible research worker: Mr. Jon NICHOL).

   Aims: -   the development of conceptual understanding in pupils;

            -    the fostering of metacognition which subsumes the crucial area of metalinguistics.

SCOTTISH COUNCIL FOR EDUCATIONAL TECHNOLOGY (SCET)
Dowanhill
74 Victoria Crescent Road
GB-GLASGOW G12 9JN

Research workers: Ms. K.F. HENNING, Mr. R.N. TUCKER

1. Annotation of software information

   In providing guidance on the selection of learning materials the project aims to maximise the effective use of learning resources, and to stimulate and sustain interest in resource-based learning. The assessment of learning materials by teachers, subject advisers, and curriculum development groups is seen as central to this process.

Through this design of a pilot database on learning resources, the potential for computerising this information is also being investigated. The aim of this development is to explore means of proving easy access to curriculum-related information on resource material.

2.   Compilation of an interactive video register, based on a survey of all Department of Trade & Industry (DTI) funded projects on interactive video. On the basis of this survey SCET will be putting proposals to DIT.

3.   Study of developments in member countries of the International Council for Educational Media (ICEM).

YUGOSLAVIA

Ana STOJEVIC
National and University Library
Marulicev trg 21
YU-41001 ZAGREB

1.   Organisation of computerised database on educational innovations (as a part of the CODIESEE Project sponsored by UNESCO; CODIESEE stands for "Co-operation in Research and Development of Educational Innovation in South and South-East Europe").

2.   Preparation of a survey (by means of a questionnaire) concerning the use of new educational technologies in secondary school media centres in Croatia.

2.3 BIBLIOGRAPHY COMPILED FROM
NATIONAL REPORTS

by
the Section for Educational Research and Documentation,
Council of Europe

UNESCO-CEPES

Centre européen pour l'enseignement supérieure, 39 rue Stirbei Vodá, R-7073 BUCAREST, Roumanie.

1. Enseignement supérieure en Europe, octobre-décembre 1985, Vol. X, No. 4.

    Numéro spécial: "L'incidence des nouvelles technologies de l'information sur l'enseignement supérieure".

2. Higher education in Europe, October-December 1985, Vol. X, No. 4

    Special issue: "Impacts of New Information Technologies on Higher Education".

COUNCIL OF EUROPE

1. New Technologies in Secondary Education - A Report of the Educational Research Workshop held in Frascati (Italy), 2-5 November 1982, Lisse: Swets & Zeitlinger, ISBN 0 947833 07 2, pp.187.

2. Sciences and Computers in Primary Education - A Report of the Educational Research Workshop held in Edinburgh (Scotland), 3-6 September 1984, Edinburgh, Scottish Council for Research in Education, ISBN 0 947833 07 2, pp.187.

COMMISSION OF THE EUROPEAN COMMUNITIES

Compendium Europ Technet, New Information Technologies and Vocational Training, published by Presses Interuniversitaires Européennes in April 1986 for the Commission of the European Communities and the European Centre for Work and Society.

THE EUROPEAN INSTITUTE OF EDUCATION AND SOCIAL POLICY

c/o Université de Paris IX-Dauphine, Place du Maréchal de Lattre de Tassigny, F-75116 PARIS.

Ladislav CERYCH and Jean-Pierre JALLADE. The Coming Technological Revolution in Education - A Report on the Potential and Limitations of New Media and Information Technologies in Education, prepared for the Dutch Ministry of Education (March 1986).

WORLD FEDERATION OF TEACHERS' UNIONS

(Vice-President: Mme Michèle BACARAT, 12 Promenée Venise Gosnat, Centre Jeanne Hachette, F-94200 IVRY-SUR-SEINE)

- Le Travailleur de l'Enseignement Technique, journal destiné aux syndique de la FISE et contenant des articles sur l'informatique

- l'IRETEP (revue de l'Institut des Enseignements sur l'utilisation des nouvelles technologies et en particulier l'informatique)

par example:

- Jean-Antide HUYNCH, Défense et illustration de l'ordinateur à l'école. In: IRETEP, Quand le courant passe, No. 2.

- Jacques-André BIZET, Effets du micro-ordinateur sur le rapport à l'écrit des élèves de LEP (Hypothèses éducatives générales). In: IRETEP, No. 6.

INTERNATIONAL ASSOCIATION FOR THE EVALUATION OF EDUCATION ACHIEVEMENT

(TO Department of Education, Division of Curriculum Technology, Twente University of Technology, PO Box 217, NL-7500 AE ENSCHEDE, The Netherlands)

Editors: R. WOLF, Tj. PLOMP, W.J. PELGRUM. Computers in education-design and planning, September 1986, 30 pp.

C O U N T R I E S

DENMARK

- Jørgen Aage JENSEN, Presentation of Project: Informatics in a Computerised Learning Environment (PILE), In: Education & Computing, No. 2/1986, pp.75-79.

- Jørgen Aage JENSEN and Bent B. ANDERSEN, Logic Programming in a Computerised Environment, Institute for Informatik, Danmarks Laerer-højskole, Saertryk, No. 7, København 1986.

FRANCE

LA DOCUMENTATION FRANÇAISE, 29/31 quai Voltaire, F-75340 PARIS CEDEX

- Le micro-ordinateur en classe maternelle - quels apprentissages? 1986, 102 pp., 60 FF

- Mutations technologiques et formations - Informatique et formation, In: Cahiers français, No. 223, 1985, 42 FF

- Le jeune enfant et le micro-ordinateur, 1984, 120 pp. (out of stock)

FEDERAL REPUBLIC OF GERMANY

Hartmut BALSER, Parkstrasse 15, D-6301 POHLHEIM 6, Die Beziehungen des Schülers zu Computer- und Informatikunterricht, In: Pilotstudy "Inter-action between pupils and computers, research based on talks with pupils in the ninth grade".

KULTUSMINISTERKONFERENZ, Sekretariat der Ständigen Konferenz der Kultusminister der Länder in der Bundesrepublik Deutschland, Postfach 2240, D-5300 BONN.

"Neue Medien und moderne Technologien in der Schule", - Bericht der Kul-tusministerkonferenz, 21 February 1986. In: Veröffentlichungen der Kultus-ministerkonferenz, Bonn, March 1986.

(Publication summing up the situation and projects in the eleven countries.)

Gerhard MAYER-VORFELDER, Neue Medien und moderne Technologien in der Schule. In: Pädagogische Rundschau 40, 1986, 6, pp.649-659.

Werber RÖHRIG, Tortonastrasse 14, D-6290 WEILBURG. Aktive Lernformen - eine Antwort der Schule auf die Herausforderung durch die neuen Medien und Technologien? In: Die Deutsche Schule 77, 1985, 5, pp.367-404.

Hans-Peter RUST, Danziger Strasse 26, D-6301 POHLHEIM, Informatik in der Sekundarstufe I - Erste Erfahrungen. In: Die Deutsche Schule 77, 1985, 5, pp.367-404.

(Evaluation of experience with computer users in lower secondary education in 37 schools of Hesse.)

VEREIN ZUR FÖRDERUNG DER PÄDAGOGIK DER INFORMATIONSTECHNO-LOGIEN, Südstrasse 135, D-5300 BONN 2.

Compass, Computer an Sonderschulen und sozialpädagogischen Berufs-bildungsstätten (specialised periodical for the use of computers in special education and the training of persons responsible for youth programmes).

NETHERLANDS

A. ANTHONISSEN, Video impuls onderwijs met computer (COO), AV-maga-zine, Vol. 7, No. 1, January 1986, pp.44-45.

Y.F. BARNARD, G. ERKENS, G. KANSELAAR, Probleemoplossing tenten-taak: analyse, simulatie en generaliseerbaarheid: dialoog structuur analyse bij interactieve probleemoplossing, Rijksuniversiteit Utrecht, Vakgroep On-derwijskunde, Utrecht, 1986, DSA-Rapport 1, Dialoog Structuur Analyse bij Interactieve Probleemoplossing.

M. BEEREN, Compact-disc basis instructief-programma, AV-magazine, Vol. 7, No. 3, March 1985, pp.60-61.

D.J. BIERMAN, P.A. KAMSTEEG, Ontwikkelingen en problemen van kennis-gestuurde onderwijssystemen. Lisse: Swets & Zeitlinger, 1985, In: Techno-logie in het onderwijs, Bijdragen tot de Onderwijsresearchdagen, H.J. BREIMER, E.J.W. VAN HEES, 1984, No. 2, ISBN 90-265-0572-8.

J. GULMANS, Begripsvorming en simulatie met behulp van computeronder-steunend onderwijs. Lisse: Swets & Zeitlinger, 1986, pp.85-95, In: Media in het onderwijs, P.W. VERHAGEN (editor), B.J. WIELINGA (editor), Bij-dragen aan de onderwijsresearch, No. 4, ISBN 90-265-0654-6.

H.A. OTTEN, Zelfsturend interactief leren met een microcomputersysteem. Lisse: Swets & Zeitlinger, 1985, pp.69-79, In: Technologie in het onderwijs, H.J. BREIMER (editor), E.J.M. VAN HEES (editor), Bijdragen tot de on-derwijsresearchdagen, 1984, No. 2, ISBN 90-265-0572-8.

SWITZERLAND

DEPARTMENT DE L'INSTRUCTION PUBLIQUE, Service de la recherche pé-dagogique, 11 rue Sillem, CH-1207 GENÈVE, Un ordinateur génevois. In: PRIM ORDI, No. 1, février 1987.

Marcel L. GOLDSCMID, EPFL-CPD, Centre est, CH-1015 LAUSANNE, Inter-face homme/Machine dans l'enseignement de demain, Publication No. 177, Ecole Polytechnique fédérale de Lausanne (Chaire de Pédagogie et Didac-tique)

Miri HALPERIN et Gabriel CHARMILLOT, Département de l'Instruction pu-blique, Service de la Recherche pédagogique, Genève, No. 87.01, janvier 1987

Raymond HUTIN, Voies et moyens, In: Information et Innovation en Education (publié par UNESCO-Bureau International de l'Education, Genève)

Jean-Marc JAEGGI et Henri SCHAERER, Utiliser le Logo en 5P, Département de l'Instruction publique, Service de la Recherche Pédagogique, Genève, No. 87.02, février 1987.

François ROBERT, Margret RIHS-MIDDEL, Marcel L. GOLDSCHMID, Jan ROZMUSKI, Serge Rochat, EPFEL-CPO, Centre-est, CH-1015 LAUSANNE, Modèle de performance, processus d'étude et compétence scientifique, Publication No. 180, Ecole Polytechnique Fédérale de Lausanne (Chaire de Pédagogie et Didactique)

François ROBERT, Serge ROCHAT, Marcel L. GOLDSCHMID, Jan ROZMUSKI, Strategies and principles for the creation of software in CAI. In: European Journal of Engineering, Vol. II, No. 2,1986, pp.135-146

Serge ROCHAT, EPFL-CPD, Centre-est, CH-1015 LAUSANNE, Les moyens futurs. Publication No. 184, Ecole Polytechnique Fédérale de Lausanne (Chaire de pédagogie et didactique)

UNITED KINGDOM

BLACKWELL SCIENTIFIC PUBLICATIONS, PO Box 88, GB-OXFORD, Journal of Computer-Assisted Learning.

Frances BLOW and Alaric DICKINSON (Eds.), The Historical Association, 59a Kensington Park Road, GB-LONDON SE11 4JH, New History and New Technology - Present into Future

Tom CONLON, Prolog brings logic into CAL implementation, May 1986. Paper presented at the EURIT '86 Conference, University of Twente, Enschede, Netherlands, 20 May 1986 (copies may be obtained from Computer Education Department, Moray House College of Education, Holyrood Road, GB-EDINBURGH

Tom CONLON, Optimism and Reality with knowledge-based CAL, July 1986. Paper presented at the PEG First Annual Conference, University of Exeter, 8 July 1986 (copies may be obtained from the address mentioned above)

Tom CONLON, Who's Afraid of Parallel Logic? July 1986. Paper presented at the PEG First National Conference, University of Exeter, 8 July 1986 (copies may be obtained from the address mentioned above)

ECONOMIC AND SOCIAL RESEARCH COUNCIL, 16 Great Portland Street, GB-LONDON W1N 6BA, Research in progress, May 1986. Occasional paper ITE/10/86, details on research supported by ESRC in the field of information, technology and education

John FOSTER, Microelectronics Education Support Unit, Advanced Technology Building, Science Part, University of Warwick, GB-COVENTRY CV4 7EZ. Booklets for various subjects in the humanities

Leslie HILLS, Interactive video in education. A report of short-term projects surveying the application of Interactive Video in educational contexts. Scottish Council for Educational Technology, 74 Victoria Crescent Road, GB-GLASGOW G12 9JN, 1986, 50 pp. ISBN 0-86011-114-8.

MICROELECTRONICS EDUCATION PROGRAMME. Microelectronics Education Support Unit, 4 Coleshill Terrace, GB-LLANELLI/Dyfed SA15 3DB. Call in the expert! A collection of case studies involving the use of expert systems in schools and colleges, 1986, ISBN 1-85126-066-8.

SCOTTISH COUNCIL FOR EDUCATIONAL TECHNOLOGY (SCET), 74 Victoria Crescent Road, GB-GLASGOW G12 9JN, Interactive Video Register, 1986.

2.4 FAMILIARISING TEACHERS WITH NEW TECHNOLOGIES

by
Dr. HANS SCHACHL,
Pedagogical Academy of the Diocese of Linz, Austria

2.4.1   INTRODUCTION

In education everything depends on the competence and the will of teachers. However, the question arises, whether the applications resulting from the new technologies will really become effective!
I want to focus upon familiarising teachers with the new developments, both in initial and in in-service teacher education and training. Research with regard to interactive learning and new technologies must be evaluated in classroom-"reality". To improve conditions for classroom-reality is the aim of every pedagogic institution. We, that is the Pedagogic Academy of the Diocese of Linz in Upper Austria, where the so-called IST-Centre ("Information - Schooling - and Training" - Zentrum) is situated, also regard this as our main concern.
The intention of this paper is to describe and give information about the activities in the IST-Centre.

2.4.2   THE IST-CENTRE IN UPPER AUSTRIA

In Austria they started with central information courses in Vienna (for grammar school teachers). In recent years five decentralised training centres have been built. The tasks of the IST-Centres are (from EDP/Informatics in Austrian Education, p. 30):
"The centre should enable all teachers to test, enlarge and increase their knowledge of appliances of various producers.
The EDP-centre should also be used for school-external adult education. There are regular courses of the 'Volkshochschulen' (public adult education centres) and working groups.
Moreover, the endeavour to open the institution to a wide public in the form of regular 'Tag der offenen Tür' should be pointed out. This 'Open House' is, above all, attended by pupils who want to put into practice the knowledge gained in computer studies and EDP-instruction. They want to work autonomously, but can be assisted by the tutors present."

In addition to that, a permanent support is being offered to teachers: A magazine library and a collection of suitable software, which partially has been developed by engaged teachers, have been introduced.
The IST-centre is situated in the house of the Pedagogical Academy of the Diocese of Linz (one of the two teacher training colleges in Linz). Therefore, the centre is used for lectures and exercises: All teacher - students

(both for primary and secondary school) have to take a basic course in computer studies at the beginning of their studies. They can even choose a special study in order to become qualified for teaching computer studies as a subject.

So, the IST-centre has become a successful institution for initial and in-service teacher training.

However, considering the fact that computer studies are, or rather should be, a general principle of education, and aim at integrating the computer into the various subjects at school, we have to increase the number of teachers to be trained. So decentralisation of training courses becomes inevitable. In Upper Austria this is carried out by the Pedagogical Institute, which is responsible for in-service training of teachers. Some of the teachers, who were trained at the IST-Centre in Linz, act as tutors and reporters in decentralised regional courses. This "multiplicator-system" should cause a kind of chain-reaction, that is to enable many teachers to use the computer for their own teaching subjects.

Future plans: To establish "mini-IST's" in the different regions of Upper Austria, where a meeting for exchange of information is to take place once a month. We even think of possibilities such as downloading suitable software from the central IST-Centre "on line".

2.4.3   PERSPECTIVES

Initial and in-service training of teachers has to consider the role of the computer in our schools in the future. Before going into detail, two questions should be answered:

First of all, which are the principal tasks of our schools? Secondly, are there any problems which can be solved in a better way, i.e. more economically, more effectively by the computer?

Schools should enable our pupils to cope with future life. Thus, the will for permanent education is necessary. The ability to develop problem-solving strategies becomes more and more important. In my opinion, a dramatic change of the curricula for several subjects has to be brought about.

2.4.3.1   *Social learning*

I wish to emphasize that computer-assisted instruction would be marked by a different teacher-pupil relationship. In the didactic remarks of the informatics syllabus, teamwork is stressed as the primary form of instruction. The computer offers ideal possibilities, especially for the so-called "project-instruction".

By portioning the different tasks, the computer enables us to solve more complex problems. At the same time, "social learning" comes in! Let me give an example: a LOGO-course was offered to parents, children and teachers in Vienna. The participants had to carry out a project (designing a meadow with flowers, a house and a sun), and each of them had to solve part of the more complex task (see Neuwirth et al., 1987, p. 230-233).

2.4.3.2 *Thinking in "networks"*

Vester pointed out that one of the reasons for the great problems of our time is the lack of thinking in networks (Vester, 1984, p. 66). Therefore, it is necessary to learn by making connections. The computer can be a valuable instrument in this task. So "project work" with computers allows and encourages co-operation between various subjects and even between different schools. Especially the "simulation programmes" are to be stressed in this context. The modern paradigm of problem-solving research is to simulate very complex situations!

2.4.3.3 *Problem-solving*

The "Lohhausen-study" (Dörner, 1983) - in my opinion an outstanding piece of work - offers a number of perspectives for an alternative policy, and above all, for a new way of learning. We have to develop the problem-solving capacities of our children by means of computers, in order to cope with the problems which we - the "sorcerer's apprentices of the evolution" - have caused!

The syllabus for computer studies in Austria includes the aim that pupils are to be conducted to a systematic way of thinking and problem solving.
"The test programme for computer studies in lower secondary schools is frequently considered as a chance to stir up the enthusiasm for problem-solving inherent in young people, and to reinforce it by means of a wider range of possible computer applications." (EDP-Informatics in Austrian Education, p. 37).
The increase of problem-solving abilities should be based upon the theoretical framework of problem-solving research (see Dörner, 1983). Again, LOGO turns out to be a possible means for learning and training of problem-solving strategies.

2.4.4 CONCLUSION

In Austria the tendency to establish computer studies as a special subject in almost all types of schools, becomes stronger and stronger. However, in order to give every individual a minimum of "computer literacy", a school should not teach a b o u t computers, but they should mainly teach by u s i n g computers. (See OECD/CERI: International Conference on Education and New Information Technologies, 1984; quoted in Bildungsinformation, 12, 1984, p. 20.)

It is the future aim of Austrian schools to integrate the computer into all subjects (see EDP/Informatics in Austrian Education, p. 33). Whether that happens or not, depends largely on the will and competence of the teachers, and so on the effectiveness of initial and in-service teacher training, as well as on the research which has to offer much practical material, and, last but not least, on the basis of educational policy: "It is necessary to have a dramatic change in the curriculum, indeed sometimes even in the fundamental nature of the various subjects." (KNIERZINGER, 1986, p. 4).

2.4.4 BIBLIOGRAPHY

DÖRNER, D. et al. (1983). Lohhausen. Vom Umgang mit Unbestimmtheit und Komplexität. Bern: Huber.

EDP/Informatics in Austrian Education. An initiative of the Federal Ministry of Education, Arts and Sports. Vienna.

KNIERZINGER, A. (1986). Problems of integrating the computer in various subjects at school. Unpublished manuscript of a lecture given at Plovdiv. Linz, Pädagogische Akademie der Diözese.

NEUWIRTH, A., TANZER, N., RITTER-BERLACH, G. (1987). Eine Studie zur Einsatzmöglichkeit der kindgerechten Programmiersprache LOGO im Unterricht. In: Erziehung und Unterricht, 4.

OECD/CERI (1984). International Conference on Education and New Information Technologies 1984. In: Bildungsinformation, 12.

VESTER, F. (1984). Neuland des Denkens. Vom technokratischen zum kybernetischen Zeitalter. München, dtv.

WEERT, T. VON (Ed.). A model syllabus for literacy in Information Technology for all teachers. ATEE, Brussels.

2.5 INTERACTIVE VIDEO PROJECTS IN FINLAND

by
OLAVI NÖJD,
University of Jyväskylä, Finland

2.5.1   INTERACTIVE VIDEO IN TEACHER TRAINING

The working group set up by the Ministry of Education for developing new teaching methods and materials for new media in education, has prepared a plan for the use of interactive video in teacher training.

The working group aims at producing an interactive video disc programme for giving information on media education in teacher training institutes in this country and possibly in other countries.

2.5.1.1 *Main objectives of the project*

To develop software for the interactive video, and explore the impact of the interactive video programmes on the learning process.

The software will consist of text units, film units, still picture units, and sound units, as well as the combination of these units. The Technical Research Centre of the State will plan the hardware combinations and computer programming. The main idea is to combine components, for example, microcomputer, touch screen, videodisc player, etc..

2.5.1.2 *Content of the programme*

The programme has been planned for teacher training institutes and teachers' in-service training. The programme aims at presenting media education curriculum in schools, teaching methods and materials, and evaluation methods using the capabilities of the interactive video.

The videodisc will be a storage for samples of different types of video programmes, samples of different teaching methods. In the videodisc, data-files of different kinds of working modes, the glossary of media education, the possibility to apply different didactical principles in media education, etc., can be found. The data-files for integration of media education into different subject areas will be of the utmost importance.

The main purpose of the programme will be to give information on media education for teachers, and to give the teachers a tool for planning, implementing and evaluating their own media education in the schools.

This project will be carried out during the years 1987-1991. The steering group "New media in education" at the Ministry of Education has been recommended to include the project in the next annual State budget.

2.5.2   INTERACTIVE VIDEO IN ART EDUCATION

The Institute for Educational Research at Jyväskylä University has been conducting research on the integration of art education to the different subject areas in the Finnish comprehensive school. Large amounts of visual information material have been collected in the form of colour slides and videotapes.

The study on the role of interactive video will be carried out during the academic year 1987-1988. The software will be stored in the videodisc; slides, videotapes and data-files on integration of art education into the different subject areas. The software on the videodisc will be chosen from the slides and other materials used in the study "Experimental teaching of an integrated and extended curriculum in drawing at the lower level of a comprehensive school in Jyväskylä."

2.5.2.1   *Main objectives of the project*

To develop software for the interactive video, and explore the impact of the interactive video programme on the learning process.

The software will consist of text units, still picture units, as well as a combination of these units. The videorecordings were due to be completed during 1987.

2.5.2.2   *Content of the programme*

The programme has been planned for teachers to help them in the planning of the art education curriculum, selecting teaching methods and materials, and evaluating the teaching process, using the capabilities of the interactive video.

The videodisc will be a storage for samples of different teaching methods. In the videodisc, data-files on different kinds of working modes, the glossary of art education, the integration (the table of integration) into the different subjects areas, can be found.

The main purpose of the programme will be to give the teachers a tool for planning, implementing and evaluating their own art education programme.

The project aims at the development of new teaching material, for use in Finland and possibly in other countries.

2.6 RESEARCH AND DEVELOPMENT OF INTERACTIVE LEARNING WITH NEW TECHNOLOGIES IN THE FEDERAL REPUBLIC OF GERMANY: MAINLY ACTIVITIES ASSOCIATED WITH THE INSTITUTE FOR SCIENCE EDUCATION

by

Dr. ROLAND LAUTERBACH,
Institute for Science Education, Kiel

2.6.1

Since 1984, increased attention has been given to the use of computers in education. During 1986, the last of the eleven states (Länder) of the Federal Republic of Germany programmatically declared the introduction of a "Basic Education in Information Technology" (Informationstechnische Grundbildung) into Secondary Education for all children, starting at grade level 8 (age 14). Materials are being developed and teachers are being trained. Although these programmes cannot be considered research, some of them are systematically evaluated in order to identify an adequate format introducing the new requirements. These are interactive, in so far as they intend the direct use of new technologies as part of an informative, qualifying and reflective introduction (for an overview see BOSLER, 1986).

Interactivity in learning has been a fundamental principle in education. Its present consideration is simply the pedagogical response to the new technical and economic possibilities of actively using media like films, slides, radio, television or video by teachers and students, and controlling their use by computers. It is supported by the expectation that the effectiveness of medial teaching will finally increase. Meta-analyses of empirical research in science teaching, so far has not been able to prove media-supported teaching to be more effective than mere lecturing, its effect-size being even slightly negative for films and television, more so for videotapes and slides (WILLETT, YAMASHITA and ANDERSON, 1983), and when compared with other teaching strategies, audio-visual methods show the lowest effect, except grading (WISE and OKEY, 1983). All in all, the pedagogical expectations, that the new technologies will be of considerable value for German schools are, at present, not shared by many educationists.

Special Education is exempted from criticism. It has received special attention as there is the conviction that the new technologies could be of considerable value for handicapped children (a summary of expectations is available in HAMEYER et al., 1987).

In the Federal Republic of Germany research and development of interactive media for educational purposes has been rather low key after a phase of high interest during the 1960s. Nationally supported institutes like the Centre for Educational Technology (Bildungstechnologisches Zentrum - BZ), Wiesbaden, or the Centre for Research and Development of Objectivated Teaching and Learning (Forschungs- und Entwicklungszentrum für objektivierte Lehr- und Lernverfahren - FEOLL), Paderborn, were even closed down during the 1970s. During this decade, especially within the last three years, attention has increased again.

For Computer-Assisted Learning (CAL), or more to the point - Intelligent Tutorial Systems (ITS), new research has been supported, and in part co-ordinated by the German Research Association (Deutsche Forschungs-gemeinschaft - DFG). A loose association of researchers in this field has just been initiated by the Institut für Informatik, University of Stuttgart (R. GUNZENHÄUSER), and the Deutsches Institut für Fernstudien (DIFF), at the University of Tübingen (H. MANDL, references at this conference MANDL, FISCHER).

The production of slides and films has been taken care of by the Deutsches Institut für Film in Wissenschaft und Unterricht (FWU), München. Inter-active video has been a concern of the Institut für den Wissenschaftlichen Film, Göttingen. German television and radio, which until last year had been publicly controlled, joined with various institutes and experts to produce educational programmes. The distributon of media has been in the responsibility of state controlled and local centres of educational media (Landes- und Kreisbildstellen).

Production of educational software, so far, has been negligible. Publishing companies have been offering mainly translated or adapted programmes from Great Britain or the United States of America. Some German programmes have also been available. Their interactive standards are low. These pro-ducts were not evaluated or research-controlled during development. A selected documentation and evaluation of some 250 programmes for the sciences and informatics from a volume of 1300 has been completed by the Institute for Science Education (IPN), in Kiel last year. This evaluation was based upon an instrument which was especially designed to identify and evaluate interactivity. (The instrument and a summary of the results have been published by LAUTERBACH, 1986.) For other subjects, the State Institute for Schools and Teacher Training (LSW), in Nordrhein-Westfalen, Soest, has done the same. At present, a common system of documentation and evaluation has been agreed upon between the responsible institutes for the Länder (RIQUARTDS, TREMP, ZIEBARTH, 1986). These institutes are now also coordinating individual software development and co-operate with publishing companies and professional software producers to establish a basic repertoire of educational software.

Educational Databases or Expert Systems are not yet available for the German school system. A survey of December 1986 (HÄUBLER, TREMP, ZIEBARTH), could not identify any research directed at general education, although at the tertiary level and for industrial purposes, expert systems exist and more are being developed.

Research on effects of new technologies in the classroom has been collected and re-analysed at the IPN since 1984 (LEHMANN, LAUTERBACH, 1985). As a result of these analyses, an effect study on social behaviour in and out of school, including classroom interaction, has been started (LANG, LEH-MANN, ANDRESEN, IPN), as well as a structural analysis of effects on reality perception and experience in elementary nature and technology studies (LAUTERBACH, 1987). A field study on the use of new information and communication technology and its socialisation effect is under way at the University of Dortmund. The pilot project was started two years ago (BAUER, HÜNERT, ZIMMERMANN, 1986).

In traditional subjects, attention toward interactive learning with computers has been given mainly in mathematics (e.g. LÖTHE, 1986), early reading and writing (BRÜGGELMANN, 1986), and simulations in physics (HÄRTEL, 1985). For physics education, a promising project with advanced interactive technology is beginning at the IPN.

Dr. Hermann HÄRTEL (1987):
Application of Modern Technology to Knowledge Capture, Design and Delivery (Enhancement of Qualitative Understanding in Physics as Basis of Modern Technology).

The project has two goals:

(1) Using the didactic potential of modern workstations with their increased capability of animated graphical display and interactivity, new concepts for the representation of basic knowledge in physics - especially in electricity - will be developed. Stressing the importance of visualisation and qualitative understanding, will support physical intuition and the capability of problem solving in this field.

(2) Using modern programming tools and methods from research in Artificial Intelligence, experience will be gained and distributed concerning the possibilities of designing, developing, and delivering complex teaching and learning material. These methods can be applied to the development of any complex material, including interactive video, computer assisted learning and tutoring, where in spite of the high degree of complexity, a high degree of flexibility can be maintained from the first draft until the last version, and a high degree of adaptability to cope with fast technological changes.

The following questions will be dealt with:

(1) What are the most common discrepancies and inconsistencies among qualitative concepts offered in traditional teaching?

(2) Is it possible to develop a consistent qualitative concept for electromagnetism which could be used in a generic form to support an introductory course, and which could be developed in a consistent way when more facts and phenomena are presented, and higher levels of abstractions are sought?

(3) What are the underlying ideas and principles which are helpful in finding these discrepancies and inconsistencies, and which could be used in a constructive way for revision and further development?

2.6.2   BIBLIOGRAPHY

BAUER, K.O., HÜNERT, M., and ZIMMERMANN, P. (1986), Sozialisation via Kabel und Bildschirm? Wie Jugendliche elektronische Medien benutzen und beurteilen. In: Jahrbuch der Schulentwicklung, Band IV. Weinheim.

BOSLER, U. (1986), Informationstechnische Grundbildung. Übersicht über die Arbeiten in den Bundesländern. In: LOG IN, 6, 5/6, p. 6-11.

BRÜGGELMANN, H. (1987), Computer als Hilfe beim Lesen- und Schreiben-lernen? In: HAMEYER, U., LAUTERBACH, R., WALDOW, J. (Eds.), Computer an Sonderschulen. Weinheim/Basel: Beltz Verlag, p. 99-108.

HÄRTEL, H. (1987), A Qualitative Approach to Electricity. Palo Alto, Cal.: Xerox Corporation, and Kiel: IPN.

HÄRTEL, H. (1985), Denken Menschen anders als Automaten? In: Physica Didacta, 12, 3, p. 11-16.

HÄUBLER, P., TREMP, V., and ZIEBARTH, W. (1986), Künstliche Intelligenz und Bildung in der Bundesrepublik: eine Bestandsaufname. Kiel: IPN.

HAMEYER, U., LAUTERBACH, R., and WALDOW, J. (Eds.) (1987), Computer an Sonderschulen. Weinheim/Basel: Beltz Verlag.

LANG, M., LEHMANN, J., and ANDRESEN, D. (1987), The Effects of Using Computers on Young People in and outside of School. Project Design for an Empirical Study. Kiel: IPN (draft).

LAUTERBACH, R. (1986), Bewertung Pädagogischer Software. In: LOG IN, 6, 5/6, p. 17-33.

LAUTERBACH, R. (in press), New Meanings of Literacy. In: STANCHEV, I., HANDIEV, B. (Eds.), Children in the Information Age: Opportunities for Creativity, Innovation and new Activities. Oxford: Pergamon Press.

LEHMANN, J., and LAUTERBACH, R. (1985), Die Wirkungen des Computers in der Schule auf Wissen und Einstellungen. In: LOG IN, 5, 1, p. 24-27.

LÖTHE, H. (1987), Lernen mit LOGO. In: HAMEYER, U., LAUTERBACH, R., and WALDOW, J. (Eds.), Computer an Sonderschulen. Weinheim/Basel: Beltz Verlag, 180-195.

RIQUARTS, K., TREMP, V., and ZIEBARTH, W. (1986), Dokumentation und Bewertung Pädagogischer Software. Kiel: IPN.

WILLETT, J.B., YAMASHITA, J.J.M., and ANDERSON, R.D. (1983), A Meta-Analysis of Instructional Systems applied in Science Teaching. In: Journal for Research in Science Teaching, 20, 5, p. 405-417.

WISE, K.C., and OKEY, J.R. (1983), A Meta-Analysis of the Effects of Various Science Teaching Strategies and Achievement. In: Journal of Research in Science Teaching, 20, 5, p. 419-435.

2.7 EDUCATIONAL RESEARCH ON INTERACTIVE LEARNING AND THE NEW TECHNOLOGIES IN IRELAND

by
Dr. DAVID OWENS,
National Institute for Higher Education, Dublin

2.7.1   INTRODUCTION

Research on interactive learning and the new technologies in first and second level education in Ireland is concerned almost exclusively with the educational role of computers. A limited number of research studies on media, other than computers, have been completed. For example, an evaluation of careers videos in secondary education was conducted during 1986 (Owens, 1986). In addition, studies on a variety of media are being undertaken by graduate students completing postgraduate degrees through research. For instance, a graduate student is conducting experimental research on the instructional effectiveness of dynamic graphics in instructional video programmes.

This summary report concentrates on significant research projects related to computer-aided learning. The report summarises studies which have been completed relatively recently, or are about to be completed, and studies planned for implementation in the near future.

2.7.2   EDUCATIONAL RESEARCH RELATED TO PRIMARY EDUCATION

2.7.2.1   *Computers in primary education*

A small-scale qualitative study of computers in primary education was completed in 1985 (Smyth and Winning, 1985). This study involved case studies of computers in four schools:

- o   a 6th class in an ordinary national school
- o   a special class in an ordinary national school
- o   a school for the hearing impaired
- o   a school for the physically handicapped

It also involved a report of a discussion about computers in primary education with a group of ten primary teachers.

The results of this study indicate that computers can promote traditional and newer educational activities (such as problem solving strategies and word processing). Computer simulations are particularly useful in providing environments in which children with physical and mental handicaps may develop life skills. Computers provide a medium which can promote inter-action in the classroom. They can enhance the teacher's role as a consult-ant/facilitator. The attitudes of school principals and parents are identified as major factors in the successful deployment of computers in classrooms. The study also revealed a widespread desire for good quality educational software.

2.7.2.2 *The role and function of computers in primary education*

The most extensive research into the role and function of computers in pri-mary education, with special reference to special education, was a two-year project which was completed in Autumn 1986. This study was conducted by personnel in the Curriculum Unit of the Department of Education, in asso-ciation with 34 different types of primary schools dispersed throughout Ireland (McNamara, 1987). The range of schools participating in the re-search included single sex and mixed schools; urban and rural schools; and schools specialising in special education.

During the two-year project (September 1984 - June 1986), different as-pects of computing were undertaken by teachers in different schools. All of the children participating in the study were introduced to the new techno-logy by provision of hands-on experience. Different kinds of hardware and an array of different types of primary school educational software were used. The categories of computer work and of computer applications in the classroom included:

a.   BASIC programming

b.   An introduction to LOGO

c.   Computer-based projects

d.   Word-processing

e.   Data storage and retrieval

f.   Data communications

g.   Control technology

h.   Simulations

i.   Content free educational software

j.   Drill and practice educational software

k.   Use of peripherals

l.   Use of special equipment

Teachers participating in the study were requested to complete three kinds of reports:

-   software evaluation reports, detailing reactions and experiences of using specific software packages;

-   project reports, describing innovative classroom projects involving computers; and

-   case study reports, summarising the total experience of using computers in the classroom.

Though some of the teachers who had participated in the study were unfavourably disposed towards computers in the classroom (that is, they were critical, worried, antagonistic, or indifferent), the majority were favourably disposed towards continued use of computers as an aid to learning throughout the curriculum. Teachers reported that they found computers useful for a variety of reasons:

a.   de-mystify computers for young people

b.   promote cross-curricular work in schools

c.   help in consolidating aspects of learning

d.   promote the development of new skills

e.   motivate children to learn

f.   introduce new areas of learning

g.   encourage a more relevant education

The results of this study have been presented to the Minister for Education for consideration, but have not yet been published. The researchers who conducted the study hope that its results will motivate the government to establish a national policy on the use of computers in primary education, and encourage it to increase its commitment of finance and other resources so that a more widespread use of computers is made in primary curricular activities.

2.7.2.3   *Gifted pupils and the use of LOGO*

An investigation of the impact of LOGO on the problem-solving capabilities of gifted children in the 9 to 13 year-old group has been initiated by Dr. Sean Close of St. Patrick's College of Education, Drumcondra in Dublin.

More than 100 gifted children attend special classes organised on Saturday mornings at several centres in different parts of Ireland. Pupils learn how to programme in LOGO and use LOGO for problem-solving. The aim of the research is to determine the extent the pupils are able to transfer problem-solving skills developed using LOGO to the solution of general mathematical problems.

2.7.2.4   *Provision, utilisation, and effect of microcomputers in primary schools*

A study of the provision, utilisation, and effect of microcomputers in primary schools is expected to be completed during 1988. It is being conducted by a schools' inspector in the Department of Education.

2.7.3 EDUCATIONAL RESEARCH RELATED TO SECOND-LEVEL EDUCATION

2.7.3.1 *Survey of computers in secondary schools*

The most extensive survey of computers in secondary education in Ireland was published in 1985 (Smyth and Winning, 1985). The aim of the survey was to identify:

(a) the extent of existing computer facilities in secondary schools;

(b) the type and usage of these computer facilities;

(c) the views of schools in relation to the future use of computers in post-primary education.

As the survey is now more than two years old, the statistical results are considerably out-of-date. However, it indicated that pupils in the later years of secondary education received far more instruction/experience and time allocation to Computer Studies, than pupils in the earlier years of secondary education. Most of the schools intended to expand the role of computers in their curricular activities, and intended to acquire additional computer software.

2.7.3.2 *European studies Computer Project*

Ireland is a member of a tripartite research project which also involves Northern Ireland and Britain. Ireland's participation is funded by the Department of Education and by a research grant from the Commission of the European Communities.

During the project pupils in six schools in each of the participating countries will create databases on topics related to local and national geography and history. Participating schools will be networked, so that each school will be able to access all of the databases. To date, teachers involved in the project, have held detailed planning meetings to agree and coordinate the overall strategy for the project. The databases will be created over three years, beginning in September 1987.

2.7.3.3 *Application of PROLOG in second level education*

An investigation of the suitability of PROLOG as a language for the creation of databases for use by second level pupils is currently being funded by the Department of Education. During 1987-1988 PROLOG will be used to create a geography database. The database will be validated using Junior Cycle classes (that is, the first three to four years of secondary school). Pupil interrogation of the database will be monitored, and teachers will be shown how to create databases using MITSI PROLOG.

During a preparatory phase of this research project, a pilot database was created in 13 subject areas, including: civics; home economics; biology; Irish; French; Latin; physics; history; geography; nature studies.

2.7.3.4 *Equality of Opportunity Project*

During 1986 the Department of Education initiated an action-research project, designed to stimulate the interest, and encourage the participation of
girls in school activities related to the new technologies. This project is
similar to the GIST (Girls in Science and Technology) project, conducted in
Manchester; but is more elaborate. It involves girls in the Junior Cycle of
second-level education, that is, girls aged from 12 to 15 years.

The project will be conducted from September 1987 until June 1988. Six
instructional modules on topics related to technology will be developed and
taught in 3 schools in Dublin and 1 in Shannon, representing four types of
schools (secondary, community, comprehensive, and vocational). Two class
periods will be devoted to the project each week. Before participating in
the project, girls will complete assessment attitude tests. Girls in four
schools not involved in the study will also complete the same tests so that
there is a control group with which to compare the effects of participating
in the instructional modules.

Upon completion of all of the modules, the pupils will complete a further
range of assessment and attitude instruments. The primary aim of the
project is to ascertain the degree of attitude change, associated with completion of the specially prepared instructional modules.

The project is being funded by the Department of Education and the Commission of European Communities.

2.7.3.5 *Application of computers in teaching the sciences*

The aim of this project is fourfold:

o   to evaluate computer programmes, equipment and ancilliary materials in
    terms of their suitability for the teaching of physics, chemistry, and
    biology;

o   to identify science topics suitable for new courseware development;

o   to develop and pilot computer-aided learning (CAL) materials on a
    range of science topics;

o   to evaluate the adoption and use of these new CAL materials.

The project was begun in 1986 and is funded by the Department of Education.

2.7.3.6 *Research being conducted by the National Information Technology*
in Education Centre

In 1986 the Department of Education set up the National Information Technology in Education Centre (NITEC), at the National Institute for Higher
Education, Dublin, at the invitation of the Commission of the European

Communities. It is one of 12 such national centres set up in each of the member states of the Community. It is funded by the Department of Education and the Higher Education Authority.

The purpose of NITEC is to collect and disseminate information on hardware, courseware, software, and activities related to the new information technologies in education. It is also responsible for the development, evaluation and dissemination of software. Its activities in this field are currently taking place in collaboration with EC-funded initiatives in school software development and portability.

NITEC is currently completing a number of research projects related to computers in first and second level education. It is completing a survey of school computer facilities, and is trying to identify the criteria which would be appropriate and acceptable for use by teachers when evaluating courseware.

It is also establishing an information network which will link the Centre to schools and colleges, teachers' centres, teacher training colleges, universities, and other third level institutions, and the Department of Education. Through this network the Centre will be able to link participating institutions in Ireland with similar institutions in other member states of the European Community through their national centres. In the first instance, this network will be primarily paper-based, but will become progressively computer-based. Modems will be placed in a limited number of schools during 1987, and modem use will be monitored to identify factors which help or inhibit the installation of a more widespread computer network.

The Centre will co-operate with other research institutions in Denmark, Scotland, and Germany in the Aarhus research project, which is designed to create and evaluate a model of portability for culture-bound software (such as history and geography). Software will be developed in accordance with the model, and will be field-tested in schools of each of the participating countries. The details of this project have been agreed with the Commission of European Communities, and the researchers are now awaiting clearance for allocation of funding.

2.7.4   CONCLUSION

To date, educational research on computers in primary and secondary education has been concerned primarily with documenting teacher and pupil reactions to educational applications of computers, and with establishing pilot projects which can motivate larger numbers of schools to become involved in innovative uses of microcomputers.

The use of computers in primary and secondary education in Ireland has been increasing rapidly during the past three years. This has been a result of many factors, notably: the reduction in the cost of computer hardware; the increase in general awareness of computer applications inside and outside school; and the increased number of research projects funded by the Department of Education and international agencies such as the Commission of European Communities.

2.7.5   BIBLIOGRAPHY

OWENS, D. (1986), An Evaluation of the Careervision Videos. Dublin: Youth Employment Agency.

McKENNA, P. (1987), National Information Technology in Education Centre. Dublin: National Institute for Higher Education.

McNAMARA, S. (1987), Computers in Primary Schools. Speech delivered to the Association of Primary Teaching Sisters, April.

SMYTH, J.J., and WINNING, I. (1985), Computers in Schools: Review and Prospects. Dublin: National Board for Science and Technology.

2.8 HOW CAN WE TRANSFORM THE RESULTS OF RESEARCH INTO SCHOOL REALITY?

by
Prof. LANFRANCO ROSATI,
University of Siena, Italy

2.8.1   SUMMARY

Three problems are considered here: research on learning, teachers and the new technologies, and in-service training of elementary school teachers.

All of them are different aspects of the same problem: school, that is to say, teachers' work, cannot be separated from society, and technological progress and educational research cannot ignore the necessity of bridging the gap between theory and practice.

Therefore this paper will follow a fixed route, within which some variables may be found which have led to special research in different sectors.

It must be pointed out that the considerations illustrated in this paper, come from the experience gained at the university where in-service training is provided for teachers of nursery, elementary and lower secondary schools, in order to meet their need for being kept up-to-date and for advanced experimentation as well.

On the other hand, initial training is deeply influenced by the experiments planned by some Italian universities, aiming at defining the characteristics of a university degree for elementary school teachers, as well as a specific pedagogical curriculum.

But the reform of the upper secondary school should come first, by which the five school years considered will be divided into a common two-year period of general education and three years of special education.

Mention must be made of the involvement of the faculties of science and humanities, together with the faculties of "Magistero" (i.e., teacher education), in planning a reform which foresees university education for elementary school teachers, too, and a specific pedagogic and didactic training for both lower and upper secondary school teachers.

This means that teachers do need to know not only the subject in itself, but how to teach it, that is to say, they must know the didactics of that subject.

In-service training, instead, is a daily activity, involving some democratic school-bodies, universities, regional centres for research, and the Ministry for Public Education.

It is in this context that the problems of research can be specified and the nature of the problems described can be found.

Subsequently, the paper will present an account of the situation of research on learning, also considering the contribution of technological devices, such as closed-circuit television, which is necessary for teachers to evaluate their own work, or the computer.

The paper will also present some experiences of learning how to read or to write, as examples of children's approach to culture in its humanistic meaning; the germinative principles of disciplines will come from this, from language to history, from art to sciences, from history to religion.

In its last section, the paper will give an account of some activities in the field op updating and experimentation, which will contribute effectively to teacher in-service training.

2.8.2   EDUCATIONAL RESEARCH INTO LEARNING

School learning is generally preceded by a sort of spontaneous pre-school learning, based on home experience. Two different levels of development must be distinguished: the affective level, which comes out when adults or school ask the child to do something, and the potential level, which can emerge and determine itself as the consequence of different stimuli, and of following acknowledgements both inside and outside school. The area of potential development of a child is very large, even if it is disabled or disadvantaged, and this area is often unexplored or superficially ignored, and, as a consequence, condemned to atrophy,

As Renzo Titone said in 1981: "School must first of all provide a more in-depth knowledge of each child's possibilities, but at the same time, it must provide each child with the widest possible range of learning stimuli encouraging the hidden or scarcely emerging potentialities."

This is to enhance the idea of early teaching, starting from nursery school, and based upon the acquisition of skills resulting from specific prerequisites.

Learning is an objective, both in terms of knowledge and behaviour. It is at the same time a process and a product. Education turns around this concept in as far as it aims at clarifying how the environment must be changed to make learning more solid and effective.

There is, indeed, an educational theory dealing with the rules of how to manage learning, which also establishes criteria for the accomplishment of certain conditions of success.

"There are too many details to be taught and learnt; we need a way to transmit the most important ideas and skills, as well as the acquired characteristics of man, those which express and enlarge man powers, we need it as never before." (Bruner, 1967).

Of course, learning is not just memorising a certain amount of information, but the acquisition and assimilation of notions and skills produced by the interaction of people and environment, in particular situations which help practice and the best possible application.

Learning is a result, too, when people master certain notions, providing them with answers to personal, deeply felt problems. There is no learning without motivation.

This last objective is not seen in terms of learning, but of knowledge acquired. As Cousinet affirmed: Learning is knowing, wanting, being able.

First of all, we need to know what can be learnt, mastered and used by an adolescent developing his intelligence, will, power, and freedom of doing and planning, believing, and thinking without any possible negative conditioning.

Knowledge leads to creativity. Ignorance is opposed to creativity, since it denies any research or diverging expression or original inventiveness.

Education gives people, especially very young people, the possibility to acquire useful information and knowledge, but, even though it always aims at knowledge, it also helps the child to achieve a sort of knowledge, leading to experience which integrates and exploits knowledge. In the proposal of "a theory of education", Bruner points out four principles which are the basis of any teaching process.

Educational theory must fix the most appropriate experiences in order to generate learning predisposition, what psychologists call readiness, to establish human relationships, and to evaluate the resources of a person whose intelligence must always be kept alert and devoted to discovery, re-building, finding new ways of knowledge. In the same time such a theory will have to specify how information must be organised in order to be understood as easily as possible by pupils, according to a developmental scale of individual modalities, governed by a system of active representation, linking thinking activity to manual expression, and by the system of iconic representation, based upon the sensorial organisation by the use of summarising images, and the use of symbolic representations, mainly built on an oral language, and on any other form of communication. The learning process can only proceed in an optimal way when material and activities are presented in such a way as to build knowledge both for learning and teaching.

The characteristics of learning stimulated in this way, acquire a teaching schedule comprising a first evaluation of learning preconditions, i.e., the prerequisites which can give a picture of individual learning predispositions.

Of course, also the learning contents have to be defined according to the idea of a structure on which each learning process is based.

In order to put the abovementioned principles into practice, attention must be paid to the sequential order in which information is transmitted, considering proposals which vary from didactic units to thematic ones, both disciplinary and inter-disciplinary ones, of course.

147

An important role in the learning process is also played by the psychological conditions of support.

One might agree with Prof. Titone, when he claims that learning as a process affects the whole person, and is based upon individual will and intelligence, but also on intuition, phantasy, imagination, memory, affective capability and reasoning.

2.8.3   TECHNOLOGY AND CLASSROOM INTERACTION

Teaching passes through communication. There are two positive elements in communication: it helps the teacher to clarify for himself what he teaches, and it helps the pupil to learn and learn how to learn. In spite of the use of didactic techniques (worksheets, objective tests, and so on), the main tool for teaching is communication in as much as didactics are considered a "theory of communication".

But it is also possible to plan individualised activities and group activities by using extremely precise technological devices, and by means of closed-circuit television. In fact, each pupil can establish a very close relationship with the screen as he interacts with it.

This interactive principle applies in particular when dealing with the computer. In this case, the teacher works at the planning level of a didactic unit (which can also involve self-evaluation tests and supporting activities); he may also interfere at any moment of the learning process as a guide and supervisor, who does not intervene by negative remarks and punishment, but in order to practise evaluation as a form of "educational intelligence" aimed at the identification of obstacles preventing pupils from individualised and meaningful learning.

Although the software may not always be relevant, or there may not be enough software available yet, it may still lead pupils to reasoned activity, that is to say, not only to reactions but also proactions. The introduction of computers at school finds any educational motivation and a didactic legitimation in the interaction it can establish, at the same time allowing the teacher greater freedom to devote time to the less gifted pupils facing learning difficulties.

Interaction in the classroom, determined by the teacher-pupil and pupil-pupil relationship, can be analysed and improved so as to render teaching more efficiently. "Macroteaching" is often used, especially at experimental level to teach future teachers how to teach. A pre-recorded fragment of a lesson can be shown to the trainees, so that they can see what happens while teaching.

The analysis of the didactical process helps to clarify the stages of the process itself (explanation, description - oral test, discussion, evaluation, practice, etc.), and confirms that teaching is "an interaction process mainly based on oral communication between teacher and pupil in certain activities" (Titone, 1964). Therefore, the teacher should programme school work, create in-depth motivation in the pupil, give essential and relevant information and coordinate and foster discussion, also by means of suggestions. He must intervene in order to correct and remove eventual disturbing elements, to verify learning conditions and ways, as well as the quality and quantity of information acquired in the form of behaviour.

Based on the categories used to analyse oral communication in the classrooms, or even better, to study oral interaction in the group, the models or systems of analysing teaching behaviour were developed by M. Hughes, N.A. Flanders, A.A. Bellack, and G. De Landsheere (Ballanti, 1979). These categories are very clear, referring both to teachers and pupils.

They are built on attitudes soliciting communication on creative processes and research of what Italians call: "accettazione empatica", listening and attention.

Didactic communication may raise problems of comprehension, since description and explanation get a higher value in communicative experience. In the cognitive process the organisation of the educational environment is very important; it is characterised by the activity of the pupils in the class in such a way that the subjective sphere comes to coincide with the objective sphere. Interactive and expressive factors come to bear in the first case, meeting with the pupil's cultural and linguistic experience, while the second case deals with the teacher as a master of the teaching-learning situation.

Didactics mean communication on several occasions: when a pupil asks for motivation, when the transmission of information becomes culture, and it should be recalled that cultural transmission is not "éducation tout court", but has to be intentional.

The aims of education are derived from philosophy, religion and tradition.

Considering just one form of culture out of many, cultural transmission may involve a conceptualism which is a guarantee for mobility among social classes or within the same class.

From the teaching point of view, cultural transmission is efficient, at developmental age level, if the contents correspond to the learning capability of pupils which differs according to age and culture.

At the educational level, organising and didactic innovations are needed in order to make the official educational bodies more critical and creative.

At the individual and social level a process has begun of mutual respect and comprehension of the different cultures, which may lead to international cooperation, and which attempts to find some transcultural elements which induce men to better understanding (Laporta, 1970). Progress and civilisation result from international understanding.

The comments on the problems of didactic communication show that some of the technological tools used as teaching aids are very useful, but also that we cannot leave it with CCTV (closed-circuit television) and computers.

2.8.4 LEARNING HOW TO READ AND WRITE

Reading and writing are the main aims of primary education (at the age of 6), even in a world of pictures such as ours. Many teaching methods are used in Italy to teach children how to read and write, and the new technology will be very helpful in this areas, too.

One concept must be recalled: a pupil reads and writes what he is induced to take into consideration. No matter whether he picks up the whole message or only part of it, what is important is that he is in a position to perceive reality, as this is the first step to enter the world of culture, i.e., the world of man, marked by man's creativity in different fields: language, science, history, arts and religion. Children must recognise the symbols referring to the "cultural areas" referred to, and understand them according to the meaning that man gave them over the centuries. In this way, science, language, history, arts and religion are productions which may be enriched by original creative personal contributions.

The first means of expression of pupils will, of course, be speaking, but writing needs manual skills, resulting from painting, collages, manual work, as well as it needs analytic skills resulting from thinking: with the help of a teacher, a pupil interprets what he already knows, using different keys which are relevant to the different subjects. The concepts of the "law of relevance" he uses as keys, are at the same time the global knowledge he already knows and the first differentiation of it.

Every reproduction by the pupil, either graphic or manual, must be recorded. So tape recorders and videotape recorders may be the first form of documentation for the child if he needs further elements, and for teachers who want to train themselves by checking other teachers' work, especially in research laboratories.

The forms of verbal expression will certainly vary from pupil to pupil, as the sources of representation and reflection are different. Every culture, as poor as it may be, bears concrete evidence in the fields of language (dialects, too), science, art, history and religion which we must know in order to enrich it.

The origin of subject matter knowledge lies in these cultural forms, in these ways of expression of human life. This is another point where technology can provide children with motivation, and the use of some common mass-media can make representation easier and thus motivate learning by perceptive activities.

It may be said that the "educational relationship" in the classroom may be improved by the new technologies being used by the teacher, who also knows how they work in order to master them, so that teaching can be as effective as possible.

The forms of experimentation and updating authorised by DPR 419 become relevant in this context. They help teachers to approach technologies for teaching, and offer training possibilities in simulated situations close to reality.

Experimental research and training activities seem to be the pivots in the hypotheses being elaborated in Italy concerning the initial and further education of compulsory school teachers. In fact, research workers concerned with the development of university curricula for pre-school, primary and secondary school teachers, are defining the features of practical classroom training to be carried out in the schools, with pupils, under the supervision of university-trained tutors.

Technologies will also be efficient when applied to practical training, from television to computer, from didactic software to tests and cards, in order to give a more scientific character to the experiment which integrates a wider, more relevant cultural education, which can only be obtained at university or at special post-graduate schools.

2.8.5   SUBJECTS, PROFESSIONAL SPECIALISATIONS AND KNOWLEDGE

This section aims at clarifying the relationship between teacher education at university (the acquisition of specific expertise at the different faculties), and the way pupils come to know the cultural reality.

There is, of course, an interactive situation: the teacher's didactic competence, on the one hand, and the pupil's need for knowledge, on the other hand. This is in particular true for the different subjects to be integrated into the "unity of culture". An example may illustrate this unity: language, before being a set of sounds, words and expressions, is a human activity which allowed man to raise above mere particularity, and enter the spiritual world. Language also includes the constitutive elements of it, namely the symbols of art, religion and science. But each language has its own grammar.

There is a science grammar, and an art grammar, and another one for myth and religion. Such grammars are not the boring thing people were taught at school, but the study of the living forms of thinking and expression.

Language, art, religion , myth, science and history are symbolic forms, or better "cultural souls", the evolution, decline and fall of which can be followed or even predicted.

Man works on them, trying to give each symbol its meaning, a meaning which can be the result of personal experience and of personal interpretation of reality, so as not to miss their generative strength.

Their force of evolution, their trend to realisation, are a guarantee for mobility in knowledge, and that expresses the cultural tension fed by each of them. Culture is a continuous spiritual production, a cultural creation of man, opening to newer and newer goals and scenarios, which become available so that man can nourish himself from it and gain new forms of life and relationship.

All subjects nowadays are no longer sets of information, assembled in such a way as to puzzle children, but they are organised into well done things, clearly showing their real structure. Their number could be reduced, compared with their number in present school-curricula; one might reduce school subjects even more by accepting McKeon's idea to regroup everything under fields such as "discovery", "recovery", "communication" and "action".

In such a perspective one would only need a nucleus of subjects showing the simplest and most significant forms of general expression, be able to determine organic sets of propositions, and to identify them without renouncing the requirements of generality and precision, what one might define as "hermeneutics" to interpret books and documents, as well as facts and actions; and the "homiletics" in order to establish relationships between "universal and particular", "cause and effect"; at last "systematisation" to ensure the in-depth vision of the connections among experience, value and knowledge.

The nature of disciplines is common of course. They are produced by man, and in this respect they allow him to master the world by experience. Anyway, language, art, history and science are perspectives from which knowledge can be derived: they make an in-depth analysis possible, and, at the same time, they allow a unitary work of reconstruction, as well as the functional synthesis that is absolutely necessary to action.

An event, an object, a datum of reality, present so many facets to the eye of those who know, such as dimensions in time, space, and culture, with linguistic implications and scientific, aesthetic ones, too. This comes from their common origin and from the exigence of man for a global knowledge: single fragments are not enough.

Analytical activities are possible in knowledge, for the distinction of any form of reality in comparison with others is also what gives its connotation and what qualifies its specificity. Such a work of analysis must be carried out by thinking, without losing sight of the peculiarity of the object he knows, that is to say, the object as a whole. The building work that is peculiar to analysis will allow the aggregation of elements that is inherent to a "point of view". In teaching activities this work is made possible by the use of the "pertinence law", which distinguishes what is pertinent from what is not.

When applied to the different forms of knowledge the pertinence law will permit the distinction of what is typical of language from what is not, what is typical of history from what is not, what is typical of science from what is not typical of it.

Immersed in a symbolic world that he himself has created, man uses disciplinary models to choose what he wants to remember from what he wants to forget. That is to say, he goes beyond knowledge meant as a mere disciplinary burden.

He is much more interested in connections, relationships and rules than in superfluous, disconnected pieces of information. Evidence for this statement can be found in the experiment of the free school group-works, described by Cousinet.

The child works as a historian, as a linguist, as a scientist applying his whole potentiality on specific contents of disciplinary nature, and using language, science, art, history and religion from the point of view of their reciprocity and unity. What is most important, he changes into a little historian, a little linguist, a little scientist, as Agosti would have said, not being satisfied with gathering information, but wondering what is history, what is language, what is science. In so doing he also establishes relationships, looks for principles, refuses mistakes, so becoming a real "little epistemologist". His attitude, when facing disciplinary knowledge, is one of critical dialogue, comparison, so that he opens to collaboration and to progress as well. A critical spirit comes from this attitude which helps the pupil to find the true essence of things.

So he learns to know and to understand what is peculiar to every symbolic form, besides its language and nature. At the same time, the child will see the limits of the "points of view" that are under analysis, he will check what has got a solution or not, in order to criticise what the history of each discipline shows as being full of errors and conjectures.

As Perkinson wrote "it is not so important that students learn a subject, but that they realise that knowledge is not perfect, those who create a subject are fallible men".

In other words, elementary school children will not be in need of knowing the history of Phoenicians at the end of their school years, but how man solved the problems of navigation (Rosati, 1986). In "To a theory on education", Bruner proposed a study-course on man, and the ideas which helped him to evolve. Then through historical experience, pupils will be able to find out the great ideas which led man in the course of time.

This is knowledge, its historical substance is relevant, and pupils can use the same system of discovery, which allowed him to learn, that is to say, the comprehension of the elements of a culture which belongs to him, and that he has to recognise and build up for himself.

Knowledge if never definite yet, as philosophy of knowledge usually points out, showing that its features are "provisional" and "subjective".

Together with the historical-sociological and institutional aspects, a theory on knowledge also involves aspects of value in order to underline the "personalising" element of knowledge. Any person, on learning notions, carried out a sort of critical evaluation aimed at the goals of knowledge, in order to build up a moral conscience of his own.

This is the case when the product of man's activity gains a significance of its own: knowledge becomes consciousness, that is to say, it becomes a personal attitude to appreciate, evaluate, criticise and compare what comes into our sphere of knowledge. More than facts, in this case, a child catches the reasons therefore, the rules which run the events, the relationships which help him to use knowledge. It is only by joining knowledge and consciousness - all that we call facts, that is to say, history, language, science, art and religion - that culture takes its connotation and becomes a moral and ethical responsibility.

2.8.6 CONCLUSION

In the pedagogic-didactic analysis of the topics inherent to this paper, the idea was to show and declare that learning always means learning to learn, and especially to learn cultural content, taking its origin from humanities. Science is not in conflict with such a connotation of culture, with what we define as the "philosophy of culture", it rather shows how to meet with it, in a meeting which is never an end in itself, and which benefits by the technological tools that are available nowadays, a meeting which leads not only to knowledge, but also to social, human, civil, and moral development.

These are the trends of research in our universities, as well as in the schools, where refresher courses are used as a means, but it is basic as well when referred to teacher initial training, or in-service training.

2.8.7 BIBLIOGRAPHY

AGOSTI, M, Il systema dei Reggenti, La Scuola, Brescia.

BALLANTI, G., Il comportamento insegnante, Armando, Roma.

BRUNER, J., Verso una teoria dell'istruzione, It ed by Armando, Roma.
Il conoscere. Saggi per la mano sinistra, It ed by Armando, Roma.

COUSINET, R., Il lavoro libero per gruppi, It ed by La Nuova Italia, Firenze.

GADAMER, H., Verità e metodo, It ed by Bompiani, Milano.

LAPORTA, R. (1979), Trasmissione culturale e educazione, in Scuola e Città, No. 8, XXX, Firenze.

McKEON, R., Gli studi umanistici e la persona, It ed by Armando, Roma.

ROSATI, L., Motivi pedagogici e didattici nei nuovi programmi per la scuola elementare, ed Borla, Roma.

TITONE, R., La scuola degli apprendimenti fondamentali, in AA. VV. Oroscopo per la scuola primaria, Armando, Roma.
La psicolinguistica oggi, PAS-Verlag, Zurich.

WATKINS, K., Libertà e decisione, It ed by Armando, Roma.

2.9 THE ROLE OF THE NEW INFORMATION TECHNOLOGIES IN DEVELOPING LEARNING AND TEACHING IN SCHOOLS

by
PER LAUVÅS,
University of Oslo, Norway

2.9.1   INTRODUCTION

I have deliberately chosen the role of the sceptic at the Workshop. The choice does not reflect an antagonistic stand towards NIT in Norwegian schools. As a specialist in education, and not in computer technology, I pursue the task of questioning and scrutinising the swift recommendations made by technology specialists for the sake of the school.

The fundamental question should be dealt with first: How can NIT be applied in order to make learning better, in order to make teaching more effective, in order to improve school climate and relations? Schools will have to change to meet new challenges in a changing society. However, we know some of the shortcomings and some of the challenges. And we need to monitor the efforts of the computer technology staff who know little about schools.

In my contribution, I will just list some of the themes I would like the Workshop to address, not elaborate upon them.

2.9.2   SOME PROBLEMS FACING TODAY'S SCHOOLS

Strong traditions exist in Norwegian schools (and in the Norwegian society) as to definition of knowledge in practical terms (mainly reproduction of factual information), as to learning (memorisation of information), and as to teaching (presentation of information). When watching TV on a Saturday night, vivid illustrations of the public concept of knowledge are presented. A person is to be admired if he is capable of reproducing many facts about the civil war in the USA, Greek mythology and what not in 30 seconds.

It is quite a challenge to put more emphasis on higher levels of knowledge in schools, because so many teachers, parents and pupils are familiar with the ancient definition of knowledge only, and are generally ambiguous to understanding, application, analysis, synthesis and evaluation, to phrase the issue in the terms of Bloom et al.

In Norwegian schools, efforts to develop the schools are more successful in lower parts of the system. In the few small schools for grades 1 to 3, considerable freedom exists to develop good learning environments, compared to the rigid and stable teaching modes and patterns of most second-ary schools. And when the knowledge level of secondary school leavers (or university entrants) is questioned and claimed to be inferior to other coun-tries, the medicine is generally found in the past. It seems to be the pat-tern that each individual finds the best model at the time when (s)he was 7 to 15 years old.

2.9.3 THE SCHOOL OF TOMORROW

Nobody claims to know the future, but everybody can - and should - try to anticipate it, and take part in shaping it. It is not difficult to agree about some predictions, and harder to draw consequences from them. I will just illustrate the point:

In the information society of tomorrow, it will not be a big problem for most people to get access to information. The problem will be not to be drowned in all the information. It is necessary to train young people of today to identify important information, to seek information they need actively, and to examine and evaluate the information they get in a critical way. It is mandatory that pupils of today leave school with a basic understanding of the core subjects in order to enable them to give meaning to pieces of information and to use the information they received.

In political terms: Democracy and democratic institutions are at stake if the generations of tomorrow are not able to seek, evaluate and use information in a highly critical way, based on a thorough scientific understanding and on social responsibility.

At the same time, young generations do change faster than the adult gene-rations can cope with. It will be more and more impossible to regard pupils as obedient, well-behaving children of former middle-class families. In a fragmented, uncertain society with progressively fewer fibres in the social web surrounding the members of the society, young people will be engaged in more activities, and more learning, outside the formal socialising institu-tions (like the school), and be too busy in their efforts to create meaning where lack of meaning has developed (Ziehe, 1982), to sit quietly in the classroom because the teacher tells them to do so.

It is not mainly a matter of revising curricula. To me it is more important to reconsider traditional concepts of learning and teaching. Most important is to do away with the close linking of teaching to teacher/textbook pre-sentation, of knowledge to memorisation and of learning to reproduction.

2.9.4 MODEL POWER

A Norwegian sociologist (Stein Braaten, 1984) launched a model power theory, which deals with the consequences of unequal competencies on the communication process. He explains how laymen in computer technology can make their own choice, either accept the superiority of the computer expert, or introduce other premises in the communication in order to counter-act the "model power" of the expert.

For a long time, schools were not considered any potential market for the computer industry. During one of the ebb phases, schools suddenly became a market for the industry. At the Ministry, inspector, head and ordinary teacher level, all keepers of the green meadows, experienced the model power from eager salesmen. Parents almost panicked; if their children did not get the opportunity to work on a computer, the children would be lost forever in the harsh competition on the labour market.

2.9.5   NIT APPLICATIONS

During the first period of courting between the computer people and teachers, many of the dead and buried ideas from the previous period of educational technology some 30 years ago popped up as apparently original ideas on how to improve the schools. Various forms of information presen-tations/ drill programmes were launched as the 'new thing'. When some of us were extremely doubtful as to the possibilities at all to keep the atten-tion of young, healthy children of today, for a period of time in such "learning situations", we were told that the "pedagogical" problems related to presentation, lay-out, etc., were minor problems!

Personally, I find four potential application areas for the use of NIT in schools:

- Standard programmes (wordprocessing, spreadsheets, database, etc.)

- Specific programmes for interactive use (e.g. simulations programmes)

- Specific programmes for mainly non-interactive use (e.g. drill pro-grammes)

- Specific software for handicapped and disabled children.

In Norway, it will not be possible, to any extent, to develop specific, national software. Development costs are so high that it is simply not viable to cover much of what is needed, without hurting others parts of school finances. At the same time, the Parliament reflects the public opinion that our own language and culture should be protected in the school from too heavy foreign influence.

Specific, non-interactive software is generally not applicable in schools, and is fortunately not on the agenda for the Workshop. The specific programmes developed for deaf, blind, mentally retarded and other handicapped children are too specific to be dealt with here, although some efforts are made in my country to help these children with the new technologies. The other two categories are those of most interest.

Standard programmes

Especially, standard word processing and database programmes have been used in schools (as far as I know of, from 2nd to 9th grade). No adaptation of programmes has been necessary, and experiences seem to be positive.

Use of word processing seems to encourage pupils to write more, and better. Especially some of those who have problems in writing, benefit from the opportunity to write neat, nice-looking text, without having to go through all the trouble in drafting, rewriting, etc. When pupils share a number of terminals in a classroom (2 to each terminal), the process of writing becomes public in a productive way. In one study, carried out by some of my graduate students, some startling examples of improvement in writing were demonstrated in 6th grade, after one year of experience with a computer, and its 8 terminals in the classroom (Jamiessen & Nyhus, 1986; Andersen, Dalland & Hetty Olsen, 1986).

Database programmes also seem to have great potential in schools.

It could be interesting to discuss if some special versions of standard software could be made for school application. My reason for suggesting such a thing is briefly as follows:

From experiences in my country, it seems to be important to use NIT in order to break down traditional ways of teaching (the teacher transmitting information and close studies of textbooks). To me it is a paradox (especially in secondary schools) that young people have to comply to a undemanding ritual and to a not effective teaching pattern, as long as they are taught, while switching over to semi-competent membership of computer scientists in the computer room in the spare time, when they are not taught.

Standard programmes offer opportunities to apply NIT, without having to learn programming. From the study mentioned above, it was evident that pupils, even in 2nd grade, benefitted from the available standard programmes. Not only word processing, they made their own database of insects gathered around the school. To me, it seems interesting to develop further ways and means of facilitating the pupils' own work with the assistance of standard programmes. It is not only a matter of economy; it is more essential that young people become proactive, and not simply consumers of software produced by others.

2.9.5.2 *Specific, interactive programmes*

In Norway, there is a substantial activity in developing various forms of specific, database-like software, applying videodisc and CD-ROM. From all the activities, I know of:

- "Dataflora" is a CD-stored database, containing plants, insects and birds.

- A complete set of maps of the country is stored on CD.

- A supplementary volume of an encyclopaedia is stored on CD.

- Historical data from one region is stored on videodisc.

- A special project on the French language is carried out together with a Danish group.

- An equivalent to the Doomsday project is carried out as a Nordic co-operation project.

This kind of development work will have to continue in pace with the development of communication systems in society at large. In addition, it is imperative that schools become less dependent upon textbooks. However, I do not for a moment expect traditional textbooks to vanish; in no way is it possible that databases and other sources of information can replace the textbook.

Interactive programmes (degree of interactiveness not specified) seem to hold great potential. This is partly true as a consequence of developments in the flow of information in the society. It is, however, a more crucial aspect that such information systems not only allow students to work actively with other materials than the textbook and the teacher's presentations, but even make it necessary for the teacher to develop his/her teaching in some desirable direction.

2.9.6   BIBLIOGRAPHY

ANDERSEN, U., DALLAND, A.M., and HETTY OLSEN (1986), Data i barne-skolen. Mimeographed report. Oslo: University of Oslo, Institute for Educational Research.

BRAATEN, S. (1983), Dialogens vilkår i datasamfunnet. Oslo.

EITRHEIM Kleivan, I. (1985), Holder LOGO hva Papert lover? Mimeographed report. Oslo: University of Oslo, Institute for Educational Research.

JAMISSEN, G. (1985), EDB, "sim-sala-bim" for spesialundervisningen? Mimeographed report. Oslo: University of Oslo, Institute for Educational Research.

JAMISSEN, G., and NYHUS, L. (1986), EDB i grunnskolen. Oslo.

JENSSEN, S., EITRHEIM Kleivan, I., SVERRE, C., and WAHLSTRØM, K. (1985), Vil du vaere med så heng på. En kritisk studie omkring informasjonsteknologiens virkning på erkjennelse, likestilling og utdanning. Mimeographed report. Oslo: University of Oslo, Institue for Educational Research.

NYHUS, L. (1985), Programutforming for interaktiv video. UNIPED, 2/85, Bergen.

ZIEHE, T. (1983), Plädoyer für ungewöhnliches Lernen, Ideen zur Jugend-situation. Hamburg.

2.10 INTRODUCTION OF THE NEW INFORMATION TECHNOLOGIES IN PRIMARY AND SECONDARY EDUCATION IN PORTUGAL

Report by
the Ministry of Education, Lisbon

An experimental project in the field of the new information technologies (NIT) is being developed at national level; the MINERVA project, aiming at the:

- Introduction of the teaching of the NIT in primary and secondary education;

- Use of the NIT as teaching aids in primary and secondary education;

- Training of teachers, guidance teachers and teacher trainers for the teaching of the NIT, and their use as teaching aids.

The project was launched in 1985. Work goes on in five centres in a decentralised way involving, first of all, five universities (Coimbra, Lisbon, Braga, Porto and Aveiro).

The Centre at Lisbon is composed of three working groups who are working: one in the Faculty of Science and Technology, one in the Faculty of Science, and one in the Planning and Research Department (GEP) of the Ministry of Education and Culture.

The working groups or centres are developing research and teacher training, as well as teaching activities in primary and secondary schools of their area.

In a first stage of development, the project aims at:

- launching and following up pilot experiments;

- recommending the most adequate equipment and methodologies;

- preparing the necessary conditions for the expansion of the project.

Other than the above-mentioned five universities, some ten basic education schools (i.e. comprehensive schools covering primary and lower secondary education), and some 40 upper secondary schools are also involved in the project.

In the short- or medium-term, the expansion of the project to new working groups and to all Colleges of Education is envisaged. Pedagogic activities will gradually be expanded to all secondary schools, and to all schools offering the final years of basic education.

At the Ministry (GEP), and within the framework of the MINERVA project, a team of research workers and teachers was set up to study the conditions of using microcomputers in basic education to bring about educational innovation.

For this purpose, five primary and four preparatory schools were equipped with computers, their teachers adequately trained to plan and use computer activities, in line with the general objectives of basic education.

These activities aim at:

- stimulating the child's creativity;

- developing new learning mechanisms;

- stimulating pupil's socialisation and co-operation;

- furthering the teacher's attitude towards an easier learning by dynamising small groups and project work;

- stimulating both teachers and schools for the dynamics of change to go along with the social and technological evolution of the present world.

Accordingly, pupils will carry out activities in the following areas:

- programming in LOGO language ("Turtle Geometry");

- use of the word processor;

- use of several programmes of educational interest (games, problem-solving, assisted drawing, graphics, databases, ...)

All these activities, initiated in the school-year 1985-86 are currently being developed in the above-mentioned schools.

2.11 AN ACTION-RESEARCH PROJECT TO STIMULATE GIRLS' INTEREST IN NEW TECHNOLOGIES

by
CARMEN CANDIOTI,
New Information Technologies Programme,
Alcalá de Henares (Madrid)

2.11.1 INTRODUCTION

This paper describes an action-research project, carried out within the framework of the New Information and Communication Technologies Programme of the Spanish Ministry of Education and Science. The aim is to encourage the participation of girls in learning via the new technologies. It is assumed that girls are less motivated towards participation in new technology-related activities than boys, as a result of sex role stereotypes in society claiming masculine superiority in scientific and technical matters.

This stereotype does not only hamper girls' cognitive development, it also constitutes a serious handicap for the acquisition of intellectual and psychomotor skills, and development of self-confidence and positive attitudes towards activities traditionally considered almost exclusively masculine. In this way, girls are placed at a disadvantage in terms of career prospects in a technologically advanced society. Research evidence indicates that children take an interest in science and technology at an early age. Sex-based differences with regard to children's interest in science and technology are established, even before secondary school age. Data reveal that girls participate in computer-assisted learning (CAL) to a lower degree than boys.

In view of this situation, the main interest of this project lies in articulating strategies for changing the classroom atmosphere, and creating a learning environment favourable to girls.

Schools should provide a series of interventions to attract girls into the mainstream of scientific and technological activities, and to develop positive attitudes towards technical tools and equipment. Schools should promote classroom innovation by using the new technologies to provide equal opportunities, and to create open learning situations, so that girls are not excluded from technological areas of experience.

Through this project, the Spanish Ministry of Education contributes to the implementation of the Equal Opportunities Programme of the Directorate-General V, of the Commission of European Communities, following the 1985 Resolution of the Council (Ministers of Education).

2.11.2   THE PROJECT

The project is conducted by Paz Gastaudi and Maria José Montero, members of the NICT-Programme, co-operating with Marina Subirats of the Department of Sociology of the Autonomous University of Barcelona.

Two groups of 35 pupils, belonging to two different primary schools in Madrid, were selected to take part in the project. Both team teachers are engaged in "Atenea" and "Mercurio" projects.

Duration of the project: January 1987 - June 1988. The project includes two stages:

The first stage lasted six months, from January 1987 until June 1987. The aim was to contact selected persons, to inform them, and to make them aware of the aims of this research project, i.e. creating and maintaining a non-sexist classroom environment. Computers and video technology should be used in an innovative way, both by the teachers and the parents of both groups of pupils.

The results of this first stage will be summed up in a full report.

The second state will last from September 1987 until June 1988. Action will be taken to encourage girls' interest in the new technologies. Teachers need to be aware of the fact that the success of the project depends upon the following assumptions being taken into account:

(a)   Girls feel reluctant towards handling technical equipment, and are unwilling to develop strategies to overcome their hesitations.

(b)   Evidence shows that many girls give up learning activities, rather than learn by making mistakes.

(c)   Problem-solving by means of new technologies requires greater cognitive abilities, and this may make girls lose confidence. This happens when the new technologies are not introduced gradually, and according to the interests and motivation of girls.

Teachers and tutors are encouraged to make science and technology more attractive to girls:

(a)   Girls show more confidence when addressed and considered personally. For this reason, verbal standards have a decisive importance. "It isn't difficult", "you can do it", etc. ...

(b)   If girls are to take part in learning experiences, a more creative learning environment will have to be developed.

(c)   The best way to stimulate girls' interests is to value their participation, however modest it may be.

At this stage of the research, a change of attitude may be expected in the case of both boys and girls. To collect information about the progress made, different tools will be used:

(a) Direct observation done by teachers and tutors at the beginning and at the end of the experience for the purpose of recording girls' interests, aptitudes and behaviour.

(b) Individual formal interviews of boys and girls.

(c) Recording information about self-confidence in groups of girls and in groups of boys.

(d) Discussions and debates with female technical experts, scientists and engineers.

(e) Visits to business firms and industries, where the work of women is relevant for success or failure.

(f) Role-playing and simulation games.

2.11.3   RESEARCH METHODS

Three tools are used: observation, questionnaires, and interviews.

Observation will be made in the classroom by teachers and tutors, and recorded on videotape.

Three different types of questionnaires are used. The first has already been given to parents, pupils (boys and girls), and teachers and tutors. The purpose is to find out motivations, interests, attitudes, and expectations as regards the use of computers and video as learning aids.

The information collected in this way, and the conclusions derived from it will be summed up in the final report. Interventions to encourage girls' interests in the new technologies will be reported as well.

The results of the questionnaires will be interpreted, and a statistical analysis of the frequency of certain answers will be carried out.

At the time of writing, no results are available, since the project is still in its early stage.

2.12 NETWORKING RESEARCH FOR INFORMATION TECHNOLOGIES AND EDUCATION – A UNITED KINGDOM PROGRAMME AND ITS LINKS OVERSEAS

by
Prof. R. LEWIS,
University of Lancaster, United Kingdom

2.12.1   BACKGROUND TO THE UNITED KINGDOM (UK) PROGRAMME

The Education and Human Development Committee was established with the reorganisation of the then Social Science Research Council in May 1982. In 1984 the Council changed its name to the Economic and Social Research Council. Early in 1983 the Committee identified and circulated for discusion an initial listing of important topics which warranted expanded support or accelerated development. The broad area of Information Technology in Education occupied a prominent place in that list. The Committee emphasised its intention that research would be centred, not only on the effect on education of machines to help teach the existing curriculum, but on the development and adaptation of the curriculum to equip people, including those of school age, to deal with intelligent machines, and to prepare them for a life changed by their arrival. For example, there are questions concerning both cognitive and organisational factors which facilitate or inhibit the adoption of Information Technology in Education, and allied to these, questions around the nature, characteristics and development of information technology literacy. These initial topics remain central to the Committee's projected agenda.

Two workshops were commissioned, and detailed discussion and workshops were held in 1983. In its further considerations, the Committee was conscious of the fact that the research community is widely scattered, and has relatively few large groups of researchers. Furthermore, it recognised the importance of involving practitioners and policy makers in the development of its programme of substantive research and research-related activities, and the necessity of ensuring close collaboration with commercial organisations, such as publishers, software houses and hardware manufacturers. It was this thinking that led the Committee away from the establishment of a single new centre to the appointment of a coordinator as the focal point for the development of the initiative throughout the country.

The brief for the Coordinator includes:

-    the review, evaluation and dissemination of the recent and current activity in the field of Information Technology and Education;

-    the identification of the needs of education in relation to Information Technology;

-    the stimulation of relevant research and the formulation of research guidelines;

- the establishment and maintenance of a database of relevant work and undertaking arrangements for coordinating and networking of those active in the field including cognitive scientists, educational researchers, practitioners and policy makers.

2.12.2   EVOLVED POLICY

In reviewing the Programme after two years, the main functions identified above remain central to its policy. These functions reflect the Programme's role in providing an infrastructure for coordinated research. The priority areas originally identified remain, viz:

- information technology literacy;
- implanting innovation and teacher education;
- AI tools in CAL development.

The foci within these areas have become more clearly defined. In particular, research is being promoted into:

- classroom processes and peer interaction;
- teachers' education through their role as researchers and developers;
- intelligent advice and explanation as an adjunct to traditional CAL simulation and games;
- IT in the whole curriculum;
- new learning environments, e.g. microworlds.

Recent extentions have led to involvement in AI in training, evaluation of vocational training, and an underpinning need to bring about better understanding, and hence collaboration between cognitive scientists and educationalists.

Research training for doctoral students and teachers is also a priority.

2.12.3   MECHANISMS

In addition to promoting the direct funding on research, the Programme(*) acts through other mechanisms:

- invited seminars of two kinds; those on specific key themes and those which provide early, critical input to new research projects;

- workshops for students and teachers which provide research training or serve to bring researchers into direct contact with peers from other institutions.

---

*   Detailed papers on the ESRC-ITE Programme are available from the author at the Department of Psychology, University of Lancaster, LA1 4YF.

-    electronic services for mail, conferencing/bulletin boards and database access (databases with personal profiles and project/publication abstracts exist at the moment);

-    diffusion of research through classical papers and documents directed at specific groups of practitioners.

2.13.4   COLLABORATION

Research and development in information technology and education is distinctive in its needs for active collaboration.

-    The field is multi-disciplinary in that its community comprises those from various traditional disciplines, computer science, psychology, education, linguistics, cognitive science. Communication between such workers is not as well established as it is within the disciplines themselves.

-    Those active in the field exist in very few newly formed groups, but most are widely dispersed and often isolated.

These features of constituency and distribution characterise new fields of study, in particular those in the new technologies, where the rate of growth is large. Moreover, recent policy is changing as indicated in the recent HM Government publication entitled "Exploitable Areas of Science" (HMSO, 1986), which offers a classification of research into three categories (fundamental, strategic and applied). The unfamiliar member of the proposed triad, strategic research, is effectively defined in terms of both social, economic potential, and scientific 'timeliness'. Basically a strategic research area is one which promises significant scientific advances in the medium term and will generate new reservoirs of knowledge. To qualify as a strategic research area, however, the knowledge it generates must also offer clear opportunities for economic exploitation. Whilst fundamental research, the normal province of Research Councils, is still seen as necessary and important, it seems likely that a shift in balance towards more strategic research is on the way (if not already in motion).

There is a need to ensure that teachers become major contributors to research, and this involvement can be seen as part of their own professional development. The ultimate goals of our research can only be achieved with the active participation of experienced practitioners. An emphasis on strategic and action research fulfils the critical criteria for innovation:
-    professional development of teachers;
-    a research orientation towards achievable goals;
-    grass-roots implantation of innovation and change.

There is much to be gained from collaborative national programmes, but international or bilateral perspectives provide important cross-cultural insights. The remainder of this paper provides an outline of some actions in which the ESRC-ITE Programme is a collaborator.

167

2.12.5   EUROPEAN INITIATIVES

2.12.5.1   *Commission of the European Communities*

(a)   Establishment of a Researcher's Network

In parallel with development of New Technologies in Education, many Member States are putting into place parallel research programmes designed to lay the foundation for the long term development in this area. Those involved actively in the implementation of NITs in education, are known to seek information about research activities and results. The purpose of this project is to arrange a European Seminar on National Priorities and Profiles on Information Technology Research in Education. This seminar (held in June 1987) will produce a report on current actions, and lead to the establishment of a database containing details of people and projects, special events related to research strategies and policies within the domain of NITs in education.

(b)   Project MEDA - Methodologie d'Evaluation des Didacticiels

This project is based in Lyon within the French REREF (Réseau Européen de Recherche en Education et Formation) initiative. It has participants from Kiel, Liège, Namur, Paris, Rome and, in the UK, ESRC-ITE Programme.

The objective is to review the various methods which are used in evaluating training materials with particular reference to computer based training. It is expected that improved methods of evaluation will result from an analysis of current practice in various organisations, academic and commercial, throughout Europe. The domain of the study is the professional training and re-training of adults.

The project is short with an interim analysis to be completed by early July, and a final report by the end of October 1987. Information for the analysis will be based on two sets of questions - one, initially drafted for the group by the participants from Namur, seeks to elicit the rationale and form of evaluation schemes used by developers and evaluators of training materials. The second set of questions are aimed at the current practices of trainers in colleges, universities and companies in their methods of training analysis.

(c)   Summer Schools on intelligent tutoring systems

Based on original proposals from Centre Mondiale (since closed), and now taken up by French and West German laboratories (at the Universities of Tübingen and Le Mans), there is collaboration in planning and gaining support for a series of seminars and summer schools (1987/88). Not only will these summer schools promote collaboration between laboratories, but they will be focussed towards the research training and experience of young researchers and aim to involve practitioners in their positive guidance towards strategic research. All those involved in steering this project have a healthy scepticism of the claims offered for the substantive contribution of AI to learning.

2.12.5.2 *Bilateral Collaboration*

(a)   CNRS/ESRC Joint Programme

This programme, which commenced in 1983, is now in its third phase. This phase has two main themes:

(i) national, regional and local policies in the UK and France, in particular, the impact of CEC policies;

(ii) information technology with reference to education and training.

Both elements will include:

-    series of research seminars;
-    study visits;
-    substantive research projects.

Within the scope of the second theme, (ii) above, the topics suitable for possible collaboration emerged from a joint seminar in December 1985. These are:

-    theories of intelligent tutoring;
-    'media specific' learning objectives and developmental consequences of single and multi-media computer assisted learning;
-    the impact of new technologies on classroom processes;
-    teacher preparation.

The seminar also saw the creation of a network linking centres of expertise in AI and education as a priority. The network would allow databases to be shared and AI tools to be accessed from any other centre. It would also provide a network of centres for a coordinated programme of study visits by researchers and students.

(b)   Réseau Européen de Recherche en Education et Formation (REREF)

This is a French initiative for cooperation and interchange in Europe. The current specific action is the CEC/MEDA project (2.12.5.1 (b) above).

(c)   Acciones Integradas - Spanish-UK Government programme

Within this programme, support for initial collaboration has been received for researchers in the Universities of Madrid and Lancaster. The current approved programme has the following objectives.

The first year of exchange support will provide a basis for long term collaboration. The two teams are well established, and the exchange will provide
the opportunity to become quite specific on the research action.

The topics on which the project will focus, initially are:

- the design of artificial intelligence based models of learning;

- a study of their feasibility and their implementation.

It is then hoped to move ahead towards:

- an implementation of the models;

- the integration of the models in authoring environments.

(d)   Yugoslav-British initiative for a "Research Programme for the
          Advancement of Learning through new Technologies"

This is a programme initiated by ESRC with the collaboration of the Monte-
negrian Academy of Sciences and Arts (Division of Natural Sciences).

It aims to buil upon initial collaborating institutions:
- Centre for Multidisciplinary Studies, University of Belgrade;
- Centre for Educational Studies, King's College (KQC), London;
- Marine Biological Institute, Kotor.

Its co-directors are:

> Professor Z. Damjanovic (University of Belgrade and Montenegrian
>   Academy of Sciences and Arts),
> Professor R. Lewis (Universities of Lancaster and London, and ESRC).

The co-directors will each take responsibility for two approaches to learning
development:
- that of physiology and biophysical mechanisms (Damjanovic);
- that of educational psychology and cognitive science (Lewis).

They will build upon the existing well established co-operation between
Britain and Yugoslavia and the existing bilateral intergovernmental programme under the auspices of the British Council, in order to:

(a) share knowledge of research and development in the application of the
    information technologies to education and training, and especially those
    pertaining to intelligent knowledge-based systems, human-computer
    interaction and human factors;

(b) further research in biophysics and physiology in those aspects per-
    taining to learning mechanisms, and to use models from those disci-
    plines in order to advance our understanding of cognitive skills;

(c) stimulate collaborative programmes of research, and to encourage exchange of students and staff from the laboratories of participating institutions.

Mechanisms: To include:

1. summer schools on the applications of information technology in education for teachers, researchers, policy makers from the developing countries.

2. summer workshops on advanced techniques in the uses of IT in education for young researchers from all nations;

3. research seminars, usually timed to take place alongside summer schools and workshops.

This initial programma hopes to draw in other European collaborators sympathetic to its overall aims, and a number of major laboratories have already expressed willingness to collaborate.

2.12.6   POSTWORD

Many European agencies, and those worldwide, seek to act as foci for collaborative actions. This is to be welcomed as long as competition for status and domination does not emerge. In our initiatives we seek neither; rather, we aim to identify the gaps in research actions or information systems which, if filled, will lead to positive outcomes for education and society as a whole. Our endeavours to work in unison with the Council of Europe, with ESF, with OECD and the CEC, are towards true pan-European actions.

For example, our approaches to ESF and the CEC consider clear needs for access to information by electronic means. A number of electronic services exist - but are they accessible? can they be improved or more focussed towards IT and Education research and development? To take a specific case of the accessibility to information, we might ask if ERIC provides an adequate service to European researchers? can we improve its accessibility? Likewise for EUDISED. Can we bring these services together and within the reach of all researchers and innovative practitioners in Europe?

Where there are holes in these systems, we should seek to fill them. Our UK programme is willing to stimulate and to join in such European efforts.

2.13 PALENQUE: AN INTERACTIVE AUDIO/VIDEO RESEARCH PROTOTYPE

by
KATHLEEN S. WILSON,
Bank Street College of Education, New York

2.13.1 BACKGROUND AND DESCRIPTION

The Palenque interactive multimedia optical disc prototype is based on themes, locations, and characters from The Second Voyage of the Mimi television show, which is being produced at Bank Street College as part of the Project in Science and Mathematics Education, initially funded by the U.S. Department of Education. The Mimi project includes a multimedia package of classroom materials - including the television show, computer software, print materials, and teacher guides - which introduces science concepts to middle-school children in a motivating and "real world" way. The "science" of The Second Voyage of the Mimi is archeology; the location is the Yucatan peninsula in Mexico.

In the television show, a cast of scientists and children explore the Yucatan's ancient Maya ruins and learn about the ways in which archeologists attempt to reconstruct and better understand an ancient culture like that of the Maya. Our Palenque interactive multimedia prototype (optical disc) has incorporated this theme to the extent that the user's experience is based on a surrogate travel exploration of an ancient Maya site, Palenque, and on the perusal of a multimedia Palenque Museum database. In our prototype, several of the characters from the television show are used to introduce concepts and provide guidance. In addition, a genuine archeologist is on hand to give the user expert information about the Palenque site and about the ancient Maya.

---

*   Adapted from a paper presented at the National Videodisc Symposium for Education, Lincoln NE, November 1986.

**   This work has been a collaborative effort between Bank Street College of Education and RCA Laboratories. Funding was provided by the RCA Corporation. Major project contributors have included John Borden, Samuel Gibbon, Donna Goldstein, Joana Hattery, Jenny Howland, Richard Hendrick, Robert Mohl, Marcia Perskie, Richard Ruopp, Jeffrey Strange, George Stuart, William Tally, and Kathleen Wilson for Bank Street College of Education, and Thomas Craver, Holly Faubel, Jesse Kapili, Richard Levine, Sandra Morris, David Ripley, and Paula Zimmerman for RCA Laboratories.

2.13.2   GOALS OF THE PALENQUE PROJECT: DESIGN AS RESEARCH

The Palenque Project has been conceived of as a research and development effort, rather than as a product development effort. Toward this end, it has involved a number of researchable experiments, both educational and technical, since its inception in September 1985. In terms of pedagogy and educational philosophy, we have experimented with issues that are at the heart of much of the research and educational product development at Bank Street College. Thus, rather than trying to teach specific facts or step-by-step procedures, we have attempted to create a multimedia (text, graphics, audio, motion, video, still video) database environment for children and their families that piques curiosity and fosters self-guided exploration, information seeking, and decision making. Rather than requiring quick reaction times or attempting to reduce the interaction time required for meaningful exchanges, our interactive multimedia environment has been designed to be detailed and varied enough to encourage extended sessions with the prototype and active, thoughtful engagement on the part of the users.

In terms of production, programming, and technical research, we have conceived of the Palenque prototype as a vehicle for designing and developing an experimental, highly interactive, multimedia (optical disc) prototype, with the breadth of a product but not the depth. Our collaboration with RCA Laboratories (the David Sarnoff Research Center in Princeton, NJ) has allowed us to experiment with state-of-the-art computing technology, which has included simulations of their DVI* technology. (DVI includes a custom [VLSI] chip set that allows for integrated full-motion, full-screen digitised video, 3D motion graphics, and high-quality digitalised audio on a single compact disc, using sophisticated compression/decompression algorithms). We have attempted to incorporate into our design as many of the advanced features of this technology as possible, given the time, money, and people resources available to us. Since our users are primarily children, we have experimented with various interface issues in an attempt to make navigation around "Palenque" motivating and comprehensible for young users. To the extent possible, we have experimented with different ways to use digital multimedia technology that are not only clear and simple enough for children to use, but informative and enjoyable as well.

2.13.3   PALENQUE PROJECT DEVELOPMENT SYSTEM HARDWARE

The hardware system we used for implementing our Palenque prototype design was comprised of a Sony LDP-2000-2 laser optical disc player, an IBM PC-AT computer with extended memory, an AT&T Targa Graphics Board, an IBM monitor, and an RCA 20" stereo color television set. Our input device is a Gravis joystick. This system was used as a development

---

*   DVI (Digital Video Interactive) is copyrighted 1987 by GE/RCA.

system for applications designed for DVI technology while it was under development. DVI was first publicly announced by GE/RAC in March 1987 at the Microsoft CDROM Conference in Seattle, Washington. Palenque prototype design, production, and research, has been under way since Septmber 1985 (to the present).

2.13.4   TARGET AUDIENCE

The Palenque prototype was designed primarily for families with children in the 8- to 14-year age range. With the home environment in mind, we have attempted to make the prototype entertaining and informative for both children and their parents. Toward this end, we have incorporated into our design simulations of many types of activities that families enjoy together and learn from, such as trips to distant places; visits to museums, libraries, and zoos; viewing television and movies; and interacting with games, books, and computer simulations.

Since it cannot be assumed that outside guidance is available in the typical home setting, one of our design goals has been ease of use. In particular, we wanted the prototype to be developmentally appropriate to the interests and capabilities of children in the target audience. We have experimented with a variety of ways to make interaction with the prototype easier, including a simplified interface using joystick inputs, a variety of visual menus and icons, and increasing complexity with increased time spent with the system.

2.13.5   THE NATURE OF THE USER'S EXPERIENCE

We have incorporated into the Palenque prototype design several discrete, yet interwoven, components for children and their families to explore in any sequence and at whatever level of detail they desire. The major components of Palenque are: video overviews; surrogate or virtual travel; a textual, visual, auditory "museum" database; characters as experts and guides; simulated tools; and game-like activities.

2.13.5.1   *Video overviews*

Four short video overviews are included in the Palenque prototype. One is a general introduction to the entire prototype; the others are overviews for each of the three major components or modes of the prototype: Explore, Museum, and Game. All the overviews are linear narrative that utilise computer graphics overlays to display menus and highlight relevant areas of the screen. Their purpose is to give users a brief introduction to such things as the nature of the prototype, the characters, the types of activities possible, and use of the joystick, menus, and icons. The overviews are the least interactive part of the prototype and the most tutorial in nature. The linear play of these overviews can be interrupted at any time if the user decides to move on to the other components of the prototype.

2.13.5.2 *Virtual Travel*

Exploration, open-ended discovery, and decision making are encouraged by the virtual travel component of the Palenque prototype. In this Explore mode, users are able to take a first person point of view "walk" or "run" around the archeological site at Palenque by indicating with the joystick which direction they would like to travel and which places on the site they would like to visit. They can move forward and reverse, up and down steep Maya steps, turn left or right at indicated branch points, and turn around 180° at any point along the route. The Palenque travel route includes a Palace with an underground maze and tower; the tomb of the ancient ruler, Pacal, in the Temple of the Inscriptions; the waterfalls of the Otulum River; and the tropical rainforest that surrounds Palenque. As users explore the site, they can choose to hear various ambient (digitized) sounds, such as rainforest insects, birds, and animals, the babble of the Otulum River, and the pounding of their (the users') weary footsteps in the humid heat of the rainforest.

The Explore mode is enhanced by several options that complement the virtual travel experience. These include: information zooms, panoramic and tilted views, and a site map with a "you-are-here" indicator. The information zooms, which are accompanied with descriptive narration by an expert archeologist, allow users to "zoom" visually into selected buildings or objects of interest for close-up views. The information provided by the zooms includes names of buildings, description of various Maya hieroglyphs, stone carvings, archeological details, and various plants and animals in the rainforest.

Panoramic and tilted views are available at selected points along the virtual travel route. By pushing the joystick left or right in the desired direction of the pan, users can see 360° views around their location on the site, as if they were turning in a circle. At certain points, users can push the joystick in the forward or reverse directions to see tilted views up or down; for example, they can look up at a tower top or down at the rainforest floor. Dynamic eyes appear to "look" left, right, up, or down as the user pans. These eyes serve as a "cue" to children that, while panning, they are "looking around", not "walking around". The map option is available at all times in the Explore mode. Users can access a schematic map of the Palenque site which indicates with an arrow where they are and in which direction they are heading. As users walk around the site, this you-are-here arrow indicates any changes of direction or location by moving on the map as the user "moves" on the paths of the site. A "jump map" option allows users to select a location of interest on the you-are-here map with their cursor, and then "jump" directly to that location on the Palenque site without having to "walk" along the paths to get there.

2.13.5.3 *Palenque Museum Database*

Curiosity and information gathering are encouraged in the Palenque prototype by the Palenque Museum database of information. Users are able to browse through the theme "rooms" of the museum to learn more about things of interest they have seen or heard as they walked around the site at Palenque. The theme rooms include a Maya hieroglyph room, a Palenque history room, a Palenque map and aerial view room, and a rainforest room. Information in the museum is accessed thematically through the four theme rooms rather than via key words. For example, users can discover information about howler monkeys by first selecting the rainforest room from a graphic of the Palenque Museum facade, then exploring the rainforst room menu (a graphic of the rainforest), then selecting the canopy layer of the rainforest menu, then selecting an icon representing mammals within the canopy, and finally selecting a photograph of a howler from a selection of canopy mammal photographs.

The information stored in the multimedia museum is in a number of different formats, including video still frames, motion video, audio descriptions and music, sound effects, computer sounds, text, and computer graphics. When learning about howler monkeys, for example, the user can access still photographs of howlers as well as "movies" of howlers in motion, text describing howlers, audio sound effects of howling howlers, and audio narration about howlers by Terry, the museum's resident expert. In addition, users can make comparisons between objects in the museum by seeing, for example, different kinds of monkeys compared in side-by-side "windows".

2.15.3.5 *Three Characters*

There are three characters in the Palenque prototype: C.T., a young teenager, and Terry, a female archeologist, who are actors in Bank Street's The Second Voyage of the Mimi television show, and George Stuart, a staff archeologist for The National Geographic Society and a Maya specialist. These characters sometimes appear as full screen video motion or as "talking heads" in windows; sometimes they are heard as audio-only narration. C.T., a character whom young users can identify, plays a major role in the video introduction and overviews, serves as a surrogate question asker, and often provides "travel tips" about places on the Palenque site while the user is in Explore mode. Terry and George offer answers to C.T.'s questions and, in the Explore and Museum modes, provide expert information about the rainforest, Maya hieroglyphs, and Maya ancient history.

2.13.5.5 *Simulated Tools*

Several simulated tools are available to users of the Palenque prototype as they explore the site and probe the museum database for information. These tools include a camera, album, tape recorder, compass, and "magic flashlight". The camera, tape recorder, and album are all devices intended to encourage users to collect and save information of interest for later reference. In this way they can create a personalised record or audio-visual "story" of their experiences at Palenque. As they walk around the site, for example, users can "take pictures" of interesting temples or stone inscriptions, or record unusual sounds, like the howl of a howler monkey, and save them in their album. Once in the album, these entries can be labeled with a "soft keyboard".

The compass provides useful nagivational information to explorers as they study panoramic views or simply walk around the site. The magic flashlight allows users to do three things: "reveal", through a simulated dissolve, old photographs of Palenque buildings showing the way they looked before they were reconstructed; get "x-ray" views into the inner structures of Palenque buildings, such as hidden stairways; and "paint on" facades of buildings are they may have looked in the days of the ancient Maya. Thus, the magic flashlight helps users to see the buildings remaining on the site the way an archeologist might see them, that is, as they may have looked long ago.

2.13.5.7 *Games and Activities*

Several games and activities have been included in our prototype in an attempt to provide motivation as well as information for family users. We have tried to make our Palenque Museum similar in some ways to a hands-on children's museum by incorporating into each of the museum rooms a game-like activity for users to play. In the Glyph room, for example, users can try to put fragmented glyph panels back together, and in the rainforest room users can try to create their own "jungle symphonies". In addition, users can choose Game Mode, which is a type of archeological scavenger hunt for places and objects of interest around the site.

2.13.6   INTERFACE ISSUES

In order to make it easy for home users to navigate around and learn to use the many components of the Palenque prototype, interface issues were extremely important in our research, design, and development efforts. The main facets of our interface are the use of visual menus and dynamic icons, an ever-present text menu bar, spatian and thematic access to information, and a simple input device. Previous videodisc design and development work at Bank Street College has shown us that highly visual menus and icons are appealing to children and easy for them to understand and use.

Thus, we have incorporated into our Palenque prototype several levels of visual menus. In a sense, the virtual travel around the Palenque site is a "main menu" for Explore Mode. In this menu various information, options, and subprogrammes (such as information zooms) are distributed spatially at meaningfully appropriate locations. with Targa Tips paint system software, we have created high resolution, colour graphics which, in the form of icons, indicate where information is available. As users walk around the site, they discover information appropriate to their location; for example, they can learn about a temple while they are standing in front of it.

In the Palenque Museum database, information is distributed and accessed thematically. The menu in Museum Mode is based upon a visual metaphor of a museum facade with representative (glyph, rainforest trees, hourglass, and globe) objects for the four museum theme rooms. Users can select any of these rooms, and browse through the information stored in them. Thus, information in the museum is accessed thematically, whereas information in virtual travel is accessed spatially. Once a representative object from the museum facade is selected, users access another level of visual menu (a scene that resembles a cutaway, close-up view of the museum room selected) with objects placed in the room for further selection.

Pictographic icons are used to represent the availability of the various options and subprogrammes users can choose at any point in the programme. They are used to indicate everything from branch points in travel, to available pans and information zooms, to narration from one of the three characters. The icons always appear on a transparent icon panel at the bottom of the screen, and change as the user switches from one mode to another. The menu bar, on the other hand, is an opaque panel across the top of the screen that is always visible and always has the same six options: help, explore, museum, album, game, and exit. In order to switch modes, get help, or exit from the programme, the user simply accesses one of these six menu bar options.

Our input device is a Gravis joystick with three buttons - one at the top and two at the base. This joystick is used for navigation around both the 2D, computer-generated space with a cursor, and for navigation around the 3D video space of virtual travel: up-forward, down-reverse, left, or right. The button at the top of the joystick is used to activate the cursor and deactivate virtual travel, or vice versa. In addition, it is used to make selections after the cursor has been moved to indicate icons or images of interest. The two buttons at the base of the joystick are used to make quick 180° turn arounds and to call up the you-are-here map while traveling.

2.13.6   DESIGN INFORMED BY FORMATIVE RESEARCH

Bank Street College has long been committed to the notion of formative research as a crucial component of any educational product development project. Formative research is a process whereby various design, production, and programming components are tested by intended users at every step along the way in the product development cycle. This kind of ongoing research provides a continual feedback loop between the various members of the design and development team, researchers, and end-users.

For the Palenque prototype we have conducted an ongoing series of formative research studies at Bank Street College with children from the Bank Street School for Children and from other schools in New York City. We have tested, among other things, children's notions of space and historical time, their ability to use maps and to navigate through our virtual space, their comprehension of our interface with its visual menus and icons, the appeal of our various characters, the Palenque site, the Palenque Museum database, and their use of the multimedia information in the museum database and various simulated tools. By combining the unique resources of Bank Street College and RCA Laboratories, we have been able to create a design and interactive multimedia prototype implementation for our target audience, which has been informed by formative research, and which draws upon state-of-the-art optical disc technology.

LIST OF PARTICIPANTS

I   CHAIRMAN, RAPPORTEUR GENERAL, AND LECTURERS

Dr. J.T. GOLDSCHMEDING (Chairman), President of the Dutch Association of Scientific Film and Video (NVWFT); Director of the Audiovisual Department of the Free University of Amsterdam, Van de Boechorststraat 1, Postbus 7161, NL-1007 MC AMSTERDAM.

Prof. Dr. C.F. VAN DER KLAUW (Rapporteur Général), Oudelandseweg 66, NL-2981 BV RIDDERKERK.

Dr. Colin HARRISON, School of Education, University of Nottingham, GB-NOTTINGHAM NG7 2RD.

Ph.D. Göran NYDAHL, Havsbadsvägen 41, S-262 oo ÄNDELHOLM (Utbildningsdepartementet, Dataprogramgruppen, S-103 33 STOCKHOLM)

Dr. Peter M. FISCHER, Deutsches Institut für Fernstudien, Hauptbereich Forschung, Bei der Fruchtschranne 6, D-7400 TÜBINGEN.

Dr. Bernard DUMONT, Directeur du Programme "Technologies nouvelles et enseignement", INRP - DP5, 91, rue Gabriel Péri, F-92120 MONTROUGE.

Mr. P.W. VERHAGEN, Department of Education, University of Twente, Postbus 217, NL-7500 AE ENSCHEDE.

Dr. Jef C.M.M. MOONEN, Centre for Education and Information Technology (COI), University of Twente, Postbus 217, NL-7500 AE ENSCHEDE.

II   DELEGATES

AUSTRIA
Prof. Dr. Johann SCHACHL, Pädagogische Akademie der Diozöse Linz, Salesianumweg 3, A-3020 LINZ.

BELGIUM
Prof. Dr. Johan HEENE, Rijksuniversiteit, H. Dunantlaan 2, B-9000 GENT.

M. Dirk KOPPEN, Koningstraat 203, B-1210 BRUXELLES

DENMARK
Mr. Lief KRAGH, Informatikkonsulent, Projektsamvirket Paedagogik og Informatik, Peadagogisk Central, Roskildevej 175 A, DK-2620 ALBERTSLUND.

FINLAND
Mr. Olavi NÖJD, University of Jyväskylä, Seminaarinkatu 15, SF-40100 JYVÄSKYLÄ.

FRANCE
Dr. Bernard DUMONT, Directeur du Programme "Technologies nouvelles et enseignement" INRP - DP5, 81 rue Gabriel Péri, F-92120 MONTROUGE.

FEDERAL REPUBLIC OF GERMANY
Dr. Roland LAUTERBACH, Institut für die Pädagogik der Naturwissenschaften, Olshausenstrasse 40-60, D-2300 KIEL.

IRELAND
Dr. David OWENS, Dean, Faculty of Education Studies, National Institute for Higher Education, Glesnevin, IRL-DUBLIN 9.

ITALY
Prof. Lanfranco ROSATI, Università di Siena, Via Risorgimento 20, I-06012 CITTÀ DI CASTELLO (PERUGIA) (excused)

NETHERLANDS
Drs. L.P.M. SCHOONDERWOERD, Ministry of Education and Science, Postbus 25000, NL-2700 LZ ZOETERMEER.

Ir. Adrian J. DUERMEIJER, Inspector, Technical Co-operation and Educational Technology, Ministry of Education and Science, Oorsprongpark 4, NL-3581 ES UTRECHT.

Mr. R.N. TUCKER, Policy Adviser, Nederlands Instituut voor Audio-Visuele Media (NIAN), Postbus 63426, NL-2502 JK DEN HAAG.

Dr. J.J. BEISHUIZEN, Vakgroep Funktieleer, Faculteit der Psychologie, Vrije Universiteit, Postbus 7161, NL-1008 MC AMSTERDAM.

NORWAY
Mr. Per LAUVÅS, Senior Lecturer, Institute of Educational Research, University of Oslo, BOX 1092 Blindern, N-0317 OSLO 3.

SPAIN
Mme Isabel ALONSO, Programa de Nuevas Tecnologías de la Información y de la Comunicación, San Ignacio de Loyola S/N, E-28806 ALCALÁ DE HENARES (Madrid).

Mme Carmen CANDIOTI, Programa de Nuevas Tecnologías de la Información y de la Comunicación, San Ignacio de Loyola S/N, E-28806 ALCALÁ DE HENARES (Madrid).

SWEDEN
Ph.D. Göran NYDAHL, Utbildningsdepartementet, Dataprogramgruppen (Ministry of Education, Educational Software Group), S-10333 STOCKHOLM.

UNITED KINGDOM
Mr. A. VAN DER KUYL, Head of Interactive Video Project, Moray House College of Education, Holyrood Campus, Holyrood Road, GB-EDINBURGH EH8 8AL.

III PROJECT NO. 8 OF THE COUNCIL FOR CULTURAL CO-OPERATION (CDCC)

Mr. Gerard H. VAN DEN HOVEN, Deputy Director of Primary Education, Directorate-General for Primary Education, Ministerie van Onderwijs en Wetenschappen, Postbus 25000, NL-2700 LZ ZOETERMEER

IV OBSERVERS

UNESCO (excused)

OECD (excused)

COMMISSION OF THE EUROPEAN COMMUNITY (excused)

WORLD CONFEDERATION OF ORGANISATIONS OF THE TEACHING PRO-FESSION (WCOTP)
M. Max FERRERO, SNI-PEGC, 209 Boulevard Saint-German, F-75007 PARIS.

INTERNATIONAL FEDERATION OF SECONDARY TEACHERS (IFST)
Mr. B. WASSER, Bureau Nederlands Genootschap van Leraren, Postbus 407, NL-DORDRECHT.

Mr. A.F. RASING, Bureau Nederlands Genootschap van Leraren, Postbus 407, NL-DORDRECHT.

EUROPEAN INSTITUTE OF EDUCATION AND SOCIAL POLICY (PARIS)
(excused)

ASSOCIATION FOR INTERNATIONAL CURRICULUM DEVELOPMENT
Ms Rosalind STEELE, Association for International Curriculum Development, Boîte Postale 105, CH-1211 GENEVE 20, or 15 route des Morillons, CH-1218 GRAND-SACONNEX.

WORLDDIDAC
Mr. A. KAPPELER, Director WORLDDIDAC, Postfach 2500, CH-3001 BERN.

V   ORGANISERS

1.   DUTCH ASSOCIATION FOR SCIENTIFIC FILM AND VIDEO (NVWFT)
Dr. Jan Tijmen GOLDSCHMEDING (President of NVWFT), Director of the Audiovisual Department of the Free University of Amsterdam, Van de Boechorststraat 1, Postbus 7161, NL-1007 MC AMSTERDAM.

Drs. Ellen BRAUTIGAM, PHITEL Multimediale Producties, Prinsengracht 211 G, NL-1015 AMSTERDAM.

Mrs. Phita STERN, NVWFT-Media Manifestatie, Postbus 9550, NL-3506 GN UTRECHT.

Mrs. Alice NOOT, NVWFT-Media Manifestatie, Postbus 9550, NL-3506 GN UTRECHT.

2.   INSTITUTE FOR EDUCATIONAL RESEARCH IN THE NETHERLANDS (SVO)
Dr. J.G.L.C. LODEWIJKS, Director, Instituut voor Onderzoek van het Onderwijs (SVO), Sweelinckplein 14, NL-2517 GK DEN HAAG.

Drs. Rob VERKOEIJEN, Instituut voor Onderzoek van het Onderwijs (SVO), Sweelinckplein 14, NL-2517 GK DEN HAAG.

3.   COUNCIL OF EUROPE
Dr. Michael VORBECK, Head of the Section for Educational Research and Documentation, Council of Europe, B.P. 431 R6, F-67006 STRASBOURG CEDEX.

Mme Danièle IMBERT, Section de la Documentation et de la Recherche pédagogiques, Conseil de l'Europe, B.P. 431 R6, F-67006 STRASBOURG CEDEX.

Printed and bound by CPI Group (UK) Ltd, Croydon, CR0 4YY

30/10/2024

01781021-0001

NEW CHALLENGES FOR TEACHERS AND TEACHER EDUCATION

A report of the Fourth All-European Conference of Directors of Educational Research Institutions, Hungary, October 1986

Edited by
A. McAlpine, S. Brown, S. Lang, E. Kentley
Scottish Council for Research in Education

This Conference was organised in co-operation with the UNESCO Institute for Education in Hamburg and the Council for Cultural Co-operation of the Council of Europe. European co-operation in educational research aims at providing ministries of education with research findings so as to help them to prepare their policy decisions; co-operation should also lead to a joint European evaluation of certain educational reforms. These conferences bring together educational research workers from the countries taking part in the work (24 countries in all), as well as from the countries in Eastern Europe. The purpose is to take stock of research findings, to review trends and innovations in education in Europe and their implications for schools and teachers, and to identify appropriate actions and areas for future research. The papers of these meetings are in general published so that ministries and interested research workers, as well as a wider public (teachers, parents, press) are kept informed of the present state of research at the European level.
The theme "New challenges for teachers and teacher education with special reference to the functions and organisation of the school" reflects current priorities of the three partners involved, as well as of most European countries.

1988, 182 pp., isbn 90 265 0884 0, bound.

in USA and Canada available from:
Taylor and Francis Ltd. International Publishers
242 Cherry Street, Philadelphia PA 19106-1906, USA

also available from the booksellers.

SWETS & ZEITLINGER B.V.   PUBLISHERS

PUBLISHING SERVICE

Heereweg 347, 2161 CA Lisse, the Netherlands

READING AND WRITING SKILLS IN PRIMARY EDUCATION

A Report of the Educational Research Workshop held in Tilburg (the Netherlands) 9-12 December 1986

Edited by:
M. Young, M. Thomas, P. Munn (Scottish Council for Research in Education, Edinburgh, Scotland)
L.F.W. de Klerk (Catholic University, Tilburg, the Netherlands)

The Tilburg Workshop was one of a series of educational research meetings which have become an important element in the program of the Council for Cultural Cooperation of the Council of Europe since 1975. This book contains the reports as well as a selection of the papers of this meeting. The theme 'Reading and writing skills in primary education' was chosen because of the growing functional illiteracy among school leavers in Europe. Although schools alone may not be able to cope with this problem, it was felt that a new approach to reading and writing in primary education might help to predict reading difficulties and prevent illiteracy through special support for poor readers.

1988, 224 pp., isbn 90 265 0880 8, bound.

in USA and Canada available from:
Taylor and Francis Ltd. International Publishers
242 Cherry Street, Philadelphia PA 19106-1906, USA

also available from the booksellers.

SWETS & ZEITLINGER B.V.   PUBLISHERS

PUBLISHING SERVICE

Heereweg 347, 2161 CA Lisse, the Netherlands

REPORTS OF EUROPEAN CONTACT WORKSHOPS ORGANISED UNDER THE AUSPICES OF THE COUNCIL OF EUROPE:

Vol. 1. Strategies for Research and Development in Higher Education
N. Entwistle (ed.)
Vol. 2. Experimental Education for Pupils Aged 10-14
J. Eggleston (ed.)
Vol. 3. Research in Science Education in Europe
K. Frey, e.a. (ed.)
Vol. 4. Stimulation of Social Development in School
C.F.M. van Lieshout, e.a. (ed.)
Vol. 5. Research into Personal Development
A. Duner (ed.)
Vol. 6. Research Perspectives on the Transition from School to Work
G. Neave (ed.)
Vol. 7. Education of Migrant Workers' Children
Vol. 8. The Evaluation of In-Service Education and Training of Teachers
A. Salmon (ed.)
Vol. 9. Sex Stereotyping in Schools
Vol. 10. New Technologies in Secondary Education
Vol. 11. Renewal of Mathematics Teaching in Primary Schools
Vol. 12. Child Development at Primary School
E.A. Lunzer (ed.)
Vol. 13. Reading and Writing Skills in Primary Education
M. Young, e.a. (ed.)

in USA and Canada available from:
Taylor and Francis Ltd. International Publishers
242 Cherry Street, Philadelphia PA 19106-1906, USA

also available from the booksellers.

SWETS & ZEITLINGER B.V.

PUBLISHERS

PUBLISHING SERVICE

Heereweg 347, 2161 CA Lisse, the Netherlands